Paths of Faith
in the
Landscape of Science

Dedication

To Annette Strunz, Edith Miller, and Ellen Helmuth,
for their life companionship.

&

To New Brunswick Monthly Meeting
of the Religious Society of Friends
for its bond of community.

Paths of Faith in the Landscape of Science

Three Quakers
Check Their
Compass

George M. Strunz
Michael R. Miller
Keith Helmuth

Chapel Street Editions
Woodstock, New Brunswick

Copyright © 2014 by the authors

Published by Chapel Street Editions,
Woodstock, NB, Canada
www.chapelstreeteditions.com

Strunz, George M., 1938-, author
   Paths of faith in the landscape of science : three Quakers check their compass / George M. Strunz, Michael R. Miller, Keith Helmuth.

Includes bibliographical references.
ISBN 978-0-9936725-0-7 (pbk.)

   1. Quakers. 2. Society of Friends. 3. Religion and science.
I. Miller, Michael R., 1932-, author II. Helmuth, Keith, 1937-, author III. Title.

BX7731.3.S76 2014   289.6   C2014-905658-3

Designed by Helmuth Productions using Adobe® InDesign.®
The text is set in Adobe Caslon Pro. Designed by Carol Twombly, Adobe Caslon Pro is a revival of Caslon, based on William Caslon's original specimen pages printed between 1734 and 1770. The titles are set in Myriad Pro designed by Robert Slimbach and Carol Twombly.

Cover Illustration: "On the Cape Split Trail" - a pastel painting by George M. Strunz. Cape Split, Nova Scotia is a curving point of land that extends into the Minas Basin at the head of the Bay of Fundy.

Photograph of the authors by Brendan Helmuth.

Contents

Preface 1

Introduction - Keith Helmuth 5
- From Theology to Continuing Revelation

Chapter One - George M. Strunz 17
- Theism to a Kind of Pantheism

Chapter Two - Michael R. Miller 35
- Imagination and Belief

Chapter Three - Keith Helmuth 69
- Faith Behind Faith

Addendum 99

References 107

Acknowledgements 115

About the Authors. 117

Faith…is our life's instinctive leap toward its origin, the motion by which we acknowledge the order and harmony to which we belong.
Wendell Berry
Another Turn of the Crank[1]

…chance left free to act falls into an order as well as purpose.
Gerard Manley Hopkins
Notebook, February 24, 1873[2]

By proceeding consistently in… scientific thinking… our modern consciousness may be widened and extended to compass the mysteries of the world… It is a path leading to a spiritual contemplation of the creative archetypal word, which brings forth man and nature out of the harmony of the universal alphabet.
Theodor Schwenk
Sensitive Chaos: The Creation of Flowing Forms in Water & Air[3]

Faith is not believing without proof, but trust without reservation.
William Sloan Coffin
Credo[4]

Preface

Scientific concepts around the evolution of life and the origin of the Universe are steadily becoming more and more of a challenge to traditional religious and cultural beliefs. This challenge includes the place of humanity in the natural world, and the question of knowledge and how it is acquired.

Although the conflict between science and religion has abated in some quarters, it remains contentious in others. Some people of strong religious faith have made peace with science and some science-minded people have been willing to give religion a respected role in personal and social life. Many people, however, remain conflicted over the claims of religion and the findings of science. Still others remain entrenched in opposition to the influence of the scientific worldview. In response, some scientists have recently mounted public attacks on the regressive influence of some forms of religion.

This conflict between religion and science is primarily a phenomenon of Western Civilization, which in its religious development has spawned ways of thinking and belief that posit an ultimate truth, and then aspire to exclusive knowledge of this truth. This put Christianity on a collision course with the rise of modern science. While almost all Christian communions historically fought against the acceptance of scientific knowledge,

Preface

there was one mutation of faith that did not – The Religious Society of Friends or Quakers as the Society came to be known.

Quakerism, from its beginnings in mid-17th century England, has never been in conflict with science, largely because a central motif of Quaker religious experience has been "continuing revelation." Along with this openness to new knowledge, Quakers generally came to regard "all truth as God's truth."

This book has been composed by three authors who share this Quaker tradition, and over many years have followed the path of faith in the landscape of science. Our stories are personal and quite different. Our vocations have also been very different. But common to our perspective is an integration of science-based knowledge into an understanding of the world and life that has enriched our sense of the sacred and broadened our experience of faith.

One of us is a trained and practicing scientist who developed a career in biological chemistry related to forestry as well as university teaching. Another is an artist, a pianist, a composer of music, and a former professor of music at university. The third is a is a retired market gardener and cider maker who once taught environmental studies and social ecology and has long been active in community economic development.

The first story recounts a journey from a Quaker childhood through a sequence of careful discernment about honesty in intellectual life and religious beliefs. The second story starts out in a Catholic upbringing and then draws on the role of imagination in forming the beliefs and values that give life meaning and purpose. It delves into how our scientific knowledge now combines with the imagination to bring a sense of the Divine into human values

Preface

and ethical practice. The third story begins in a rural Mennonite childhood and an early intuitive sense of Earth's ecological reality. This leads to a recombination of science and faith, which, in turn, becomes a quest for ecological guidance.

In telling these stories we do not imagine we speak for Quakers in general, but only that the Quaker way of being in and knowing the world has been central to our journeys – a faithful compass as it were.

****

The Introduction provides a review of the historic conflict between Christian theology and the rise of modern science, and an update on how this conflict has in recent times evolved into a dialogue that has enhanced both the theological and scientific worldviews. Each chapter then provides the author's story within the context of this wider dialogue. The book concludes with an addendum that provides some historical background on the role of Quakers in science.

This book grew out of an ongoing dialogue between the authors and a common motivation that often comes with aging – the desire to do a little summing up. We hope it might inspire others who have traveled the paths of faith in the landscape of science to check their compass and tell their stories as well.

George M. Strunz
Michael R. Miller
Keith Helmuth

Introduction
From Theology to Continuing Revelation
Keith Helmuth

Part I

The realization that culture stands behind religion and that the ecology of earth stands behind culture has been a shock to the self-image of Western Civilization that is still echoing in theological and philosophical circles. In addition, the realization that truth, meaning, and guidance rest with the creation and maintenance of good stories is a further shock now percolating through both religion and science.

In the High Middle Ages, theology was the "queen of the sciences." Knowledge and guidance flowed from the Christian theological worldview. By the 17th century, Galileo was offering a path to knowledge that radically diverged from theology. He put it this way:

> *The universe cannot be read until we have learned the language and become familiar with the characters in which it is written. It is written in mathematical*

Introduction

> *language, and the letters are triangles, circles and other geometrical figures, without which means it is humanly impossible to comprehend a single word. Without these, one is wandering about in a dark labyrinth.*[1]

Although Galileo did not explicitly challenge the authority of theology and the Scholastic method of creating knowledge, the leadership of the Roman Catholic Church was sufficiently disturbed by this divergence that it famously attempted to quash Galileo and his emerging worldview.

In the early 19th century, Carl Friedrich Gauss named mathematics "queen of the sciences," and the Christian worldview was reduced to walling off theology into a realm of discourse that now ran parallel to the rapidly developing natural sciences. Not only were the various sciences beginning to create very good, — i.e., believable — stories about the structure and processes of the physical world, the translation of this increasingly precise knowledge into technology and industry was a convincing demonstration that "science" was revealing "the really real."

The Latin word translated as "science" means simply "a field of knowledge." This is why theology could once think of itself as the "queen of the sciences." But with the introduction of the empirical method by Francis Bacon in the early 17th century, and its rapid adoption, "science" came to mean the "natural sciences." Thus began the specialization of inquiry, leading to the discrete and precise fields of knowledge that seem normal to us today.

"Real" knowledge came to be seen as deriving more and more from science, and its guidance for dealing with "reality" seemed increasingly credible. In the face of this shift in consciousness, theology came to seem more and more speculative, and religious, or spiritual, knowledge became isolated into a separate and parallel worldview.

In recent times some scientists who had a feeling for the unity of knowledge, or at least for a unity of method for arriving at knowledge, began to research the significance and meaning of religion. Similarly, theologians who wanted to retain the credibility of their knowledge began to enrich their thinking with the concepts and metaphors of science.

In the last half century or so, this respectful cross-fertilization has been guided into a productive dialogue by a number of scientific thinkers who were sensitive to the persistence of religious culture, and by religious thinkers who were sensitive to the power of the cosmological and evolutionary insights of the sciences. The antecedents for this new flowering of dialogue between science and religion include Alfred North Whitehead (*Science and the Modern World*), William James (*Varieties of Religious Experience*), Arthur S. Eddington (*Science and the Unseen World*), Pierre Teilhard de Chardin (*The Phenomenon of Man*), John Dewey (*A Common Faith*), Henri Bergson (*Creative Evolution*), Paul Tillich (*The Religious Situation*), and Aldous Huxley (*The Human Situation*).

In the last thirty years the dialogue between science and religion has blossomed into a veritable garden of hybridized understandings. There are a number of figures from science that have worked out an accommodation with their religious views and have been articulate in sharing the way they understand the relationship. These include Ian Barbour (*When Science and Religion Meet*), Paul Davies (*God and the New Physics*), John Polkinghorne (*Exploring Reality: The Intertwining of Science & Religion*), Jocelyn Burnell (*A Quaker Scientist Reflects: Can a Scientist Also Be

Introduction

Religious?), George F. R. Ellis (*On the Moral Nature of the Universe: Theology, Cosmology, and Ethics*), Francis S. Collins (*The Language of God: A Scientist Presents Evidence for Belief*), and Stephen Jay Gould (*Rocks of Ages: Science and Religion in the Fullness of Life*).

From theology and religion there are thinkers who go beyond accommodation to a realization that science is unfolding a "new story" of the emergence and evolution of planetary life and that this story provides a cosmology of relationship for a rebirth of religious experience. These include Thomas Berry (*The Sacred Universe: Earth, Spirituality, and Religion in the Twenty-First Century*), Matthew Fox (*Creation Spirituality*), Sallie McFague (*Models of God: Theology for an Ecological, Nuclear Age*), Philip Hefner (*The Human Factor: Evolution, Culture, and Religion*), and John Haught (*God After Darwin: A Theology of Evolution*).

Another flowering in this cultural garden is that of natural scientists, philosophers, social scientists, and cultural historians who take religion seriously as a human phenomenon. These include David Sloan Wilson (*Darwin's Cathedral: Evolution, Religion, and the Nature of Society*), Edward O. Wilson (*The Creation: A Meeting of Science and Religion*), Stuart A. Kauffman (*Reinventing the Sacred: A New View of Science, Reason, and Religion*), Ursula Goodenough (*The Sacred Depths of Nature*), Loyal Rue (*Nature is Enough: Religious Naturalism and the Meaning of Life*), William Y. Adams (*Religion and Adaptation*), Mark C. Taylor (*After God*), Daniel Dubuisson (*The Western Construction of Religion*), Ursula Franklin (*The Ursula Franklin Reader: Pacifism as a Map*), and William Grassie (*The New Sciences of Religion*).

This is but a small sampling of the scientists and scholars of religion who have become engaged in the science and religion dialogue, and in the effort to create "a new story" in which this dialogue can flourish.

Introduction

Part II

The dialogue between science and religion is like the cat in the old song, "The Cat Came Back."²

> *But the cat came back the very next day. The cat came back. They thought he was a goner, But the cat came back. He just couldn't stay away.*

When Galileo figured out that the earth moved around the sun, the "cat" woke up. But it got a swift kick to the head from the Church fathers and collapsed in a corner. When Hutton and Lyell's work in geology began to reveal the age of the earth and the forces and processes that shaped the crust and landscape of the planet, the "cat" woke up again and began to prowl around through all of natural history. When Darwin published his research and analysis in *The Origin of Species*, the "cat" got up on its hind feet and started scratching at the front door of truth.

Serious thinkers in the field of theology in those days were quite certain that all truth was God's truth and God's truth was in the Bible. But with Hutton, Lyell, and Darwin, the "truth" about the earth, its history, and the history of life could not be squared with the Bible, and the science and religion debate was joined. This "cat" took a commanding seat at the table of cultural discourse, and for a number of decades the fur really flew.

After big arguments, personal attacks, and even legal proceedings (Scopes Monkey Trial), the "cat" was exhausted and went out for a long walk. The opponents licked their wounds and for a while carried on in their separate domains of cultural life. But "the cat came back." After the furor caused by the "death of God theology" in the early 1960s, something changed in both the culture of religion and the culture of science.

Introduction

Serious thinkers in theology who had gone through the wringer of the death-of-God debate, began to see that although certain theological concepts can be rendered unbelievable and functionally sidelined by cultural change, religion and the hunger for authentic spiritual experience was not, thereby, knocked off the rails. Theologians were not exactly unemployed, but their job description had to be rewritten.

Serious thinkers in science began to realize that although the old nemesis, institutional religion, was increasingly moribund, new forms of religious expression were rebounding and, in some cases, were incorporating science-based narratives in their worldview. This religious appropriation of science-based stories brought scientists into a new dialogue with religion. In addition, the obvious persistence and re-emergence of religious culture became an interesting new area of scientific research.

Gradually, hapless theologians began to look at the culture of science and understand that something of profound significance for the human story was coming to light in this domain. Similarly, bemused scientists began to look at the determined efforts of religious thinkers to bring the revelatory stories of science into a renovated worldview that made sense of the human situation. A common focus on the human story and its holistic context drew scientific and religious discourse more and more into mutually beneficial dialogue.

While this rapprochement began to emerge in many centres of study and research, the collaboration of Metanexus Institute and the Templeton Foundation is a particularly notable blossoming of the recent science and religion dialogue. Metanexus began as the Philadelphia Center on Religion and Science in the mid-1990s. Founded by Quaker activist/scholar, William Grassie, the Institute developed a strong association with the Templeton Foundation.

The Templeton Foundation was already a pioneer in funding research and publishing in the area of science and religion and became a major financial backer of Metanexus Institute. In addition to holding major conferences that brought hundreds of scientists and scholars of religion from around the world together for week-long sharing and dialogue, the Institute's programme also helped establish and fund science and religion research projects in a variety of locations around the world.

The wave of interest in the science and religion dialogue that Metanexus helped generate and develop has now settled widely in many venues. Interdisciplinary courses and programmes fostering the dialogue between science and religion are now found at many colleges and universities. In addition, the growth of neuro-cognitive research, evolutionary psychology, and bioethics has spurred the science and religion dialogue to develop a more unified form of guidance for understanding the human story.

This renewed science and religion dialogue has continued to blossom in ever more interesting forms right around the world. Scholars of Islam, Buddhism, and the religious cultures of India have enthusiastically rallied to the international, multi-disciplinary institutes, conferences, and colloquia that are now held around the planet. Scholars from the mainstream Christian traditions, together with scholars from other religious traditions, now regularly gather with cosmologists, physicists, biologists, ecologists, chemists, anthropologists, paleontologists, bioethicists, environmentalists, economists, computer scientists, and historians of science to share their findings about what is going on in their respective traditions of research and practice, and to ponder the meaning of it all for the human story.

Introduction

There are still disagreements, debates, and searching exchanges in this renewed dialogue, but these folks come together because they want to be together and expect to have a mutually beneficial encounter. This is a far cry from the "war between science and religion" that once consisted mostly of each side scoring points at the other's expense. The key to understanding the "cat" that has come back is a common interest in and commitment to the human story.

Dogmatic theology is now a whisper of its former self. Science, too, has come up against factors of method that stymie its expectations of certainty about many things. The "cat" has come back to the table with a good helping of humble pie all around, and the intellectual digestion of the dialogue on science and religion is much improved over former days. In the past, the two camps seemed to have little in common, but now they both know that the fate of the human story is at stake, and that they both have a role in how this story continues to unfold.

Thoughtful theologians now know that scientific knowledge has placed humanity in a "new story" about the unfolding of life within the Cosmos and about how life actually works in the biosphere. Thoughtful scientists now know that religious traditions excel at encapsulating and transmitting the master narratives of culture. The human story is in a crisis of transformation. Both science and religion are intimately and powerfully involved in this crisis. It's a good thing the "cat" has come back. The human prospect needs all the help it can get.

However, several high-profile atheists have recently thrown a curve ball into this picture. Richard Dawkins, Sam Harris (both scientists), and Christopher Hitchens (cultural analyst) have each written substantial books and engaged in public debate

Introduction

arguing against the reasonableness of religious belief, and against religion as a progressive factor in human history. They have each, in their own way, thrown down the gauntlet for reigniting the war between science and religion. They clearly want science to win and religion to wither. (Their books are cited and referenced later.) But, in looking carefully at their volleys and at the response, a different scenario seems to emerge.

The science and religion conversation has now widened to the point that even these attacks find a place in the dialogue. Try as they might, they are not hurling their brickbats from the outside. They are part of the conversation whether or not they want to be. The fact that some folks who take religion seriously find their thinking and analysis refreshingly helpful indicates the strength and resilience of the new cultural platform that is developing around the science and religion dialogue. Dawkins, Harris, and Hitchens are simply carving out a little space on the platform on which to stand.

The eminent 20th century theologian, Paul Tillich, identified religion as "being grasped by an ultimate concern," and faith as "the state of being ultimately concerned."[3] The culture of religion and the culture of science have thus become increasingly overlapping domains. Again, the key to this cultural change is a common understanding by both science and religion that what we are dealing with is an "ultimate concern" about the human story and the question of human betterment. Dawkins, Harris, and Hitchens would not have gone to the trouble of writing their books if they did not share this ultimate concern.

The upshot of this relationship was once well expressed by Edwin Markham in the poem "Outwitted,"[4] although in this case the roles are reversed.

Introduction

> He drew a circle that shut me out–
> Heretic, rebel, a thing to flout.
> But love and I had the wit to win:
> We drew a circle and took him in!

This little book follows in the tradition, dear to Quakerism, of drawing a larger circle of inclusion. *Paths of Faith in the Landscape of Science* is an expression of how the dialogue between science and religion has been working out in the lives of three Quakers for over half a century. Typical of the Quaker approach, our stories are long on experience and short on metaphysics. This puts us in good company with the "cat" that came back and is now drawing a large circle around the human story.

For the authors of this book, Quakerism has long been a religious path that is entirely at home in the landscape of science. Quakerism is unique in the heritage of Christendom in its relationship with the unfolding of scientific knowledge. Not only was the growth of scientific knowledge generally accepted by Quakers, but various Quakers even became leaders in the development of science, technology, and industry.

The Religious Society of Friends emerged in England at the same time the scientific revolution was beginning to get underway. Both movements grew from a cultural transformation that saw the medieval worldview give way to the modern worldview. The Religious Society of Friends was pivotal in realizing the potential of this change for religious culture. This was not necessarily what Quaker leadership aspired to or thought they were doing, but certain key elements in Quaker experience and discernment fostered this historic role.

Introduction

For example, the founding figure in Quakerism, George Fox, experienced an "opening" that became the basis of a spiritual method to which a large number of people responded with a sense of recognition. Fox was not the only one, nor even the first, in whom the realization arose that authentic Christianity was not found in outward forms or in particular language, but in the direct experience of God through the spirit of Christ. Following in the wake of Luther's Reformation over a century earlier, the Anabaptist movement set out to recover, as they saw it, this essential truth of the gospel message.

There is, however, in Fox's recovery of this truth, an emphasis that distinguishes it from the Anabaptists, and, indeed, from all previous Christian theology. For Fox it was not just a direct relation to God with respect to salvation, but the immediate experience of direct teaching by the inward Christ, the experience of "continuing revelation," as Quakers came to call it. His announcement that "Christ has come to teach his people himself"[5] *shifted the basis of spiritual life from a preoccupation with personal security to an engagement with the process of learning.*

Into a milieu of dogmatic theology and rigid church structure already under siege, Fox projected a new horizon of spiritual life, a new horizon of learning – the prospect of learning directly from the immediate counsel of the indwelling Christ. The Quaker mode of silent, expectant worship established a discipline of listening and learning that became a method of access to new knowledge. Growth replaced security as the dominant metaphor of religious experience. Fox opened this new horizon of learning from a biblical base, but the process, as such, is not uniquely Christian or even biblical. This openness to learning is a fundamental potential of human intelligence. Many early Friends had the gift to see that

Introduction

what they were about is universal to the species, the potential of both genders of every rank, race, culture, and creed.

Following from this spiritual and cultural innovation, Quakers have been pioneers in education and in many fields of human development and social betterment. Many Friends have been attracted to the sciences and scientists have been attracted to Quakerism. We may wonder why so much modern social analysis, so many programs of experiential learning, so many problem-solving processes, and so many contemporary programs of social action that have no direct link to the Religious Society of Friends, seem, never the less, as if they have a touch of Quaker DNA. In a real sense, they have just that.

If we study the shift in Western culture from a set worldview to an evolutionary perspective, and from the certainty of unchanging knowledge to an open horizon of learning, it is not difficult to see that the innovation in spiritual and cultural life that Quakers launched is one of the primary sources of this change. This innovation is what Kenneth Boulding called the "evolutionary potential of Quakerism,"[6] and it is why Quakers have never been thrown off stride on their path of faith by the unfolding landscape of science.

In the essays that follow, each author reflects on his own path of faith in the landscape of science, and, using the Quaker compass, looks to the direction from which "continuing revelation" may yet come.

Theism to a Kind of Pantheism
A Scientist's Personal Journey
George M. Strunz

I was raised in a Quaker family in Dublin, during the 1940s and 50s, my parents having joined the Religious Society of Friends after our family's escape from Nazi tyranny in Vienna in 1938 when I was an infant. Although Ireland has long been a predominantly Catholic country, at that time more than 1500 Irish Friends attended some 30 Quaker Meetings there.

Indeed, Quakerism has enjoyed a long and respected presence in Ireland. Both George Fox, the founder of the Society of Friends, and William Penn, Quakerism's best-known historical figure, spent significant time there. Irish Friends Meetings followed, and still adhere to, the traditional Quaker religious service based on meditation and Christ-centred silent worship.

When I was growing up, we were, naturally, considered to be part of the Protestant minority by our Catholic neighbours. In some quarters, I could feel that there was still a residue of extra good will towards Quakers resulting from Friends' relief work during the Potato Famine.

Chapter 1

Friends are not required or expected to subscribe to any creed and are further distinguished from other Protestant sects by several unique testimonies, notable among them, the Peace Testimony. Quaker philosophy includes the belief that there is something of God in everyone. A noticeable feature in Friends Meetings is the absence of outward sacraments. Meeting Houses are very simply furnished. Important as these distinguishing features are, generally speaking, Friends embraced a theology that had much in common with Protestant-Christian Faiths. The "Presence in the Midst," an engraving that adorned the wall of the lobby of the former Eustace Street Meeting House in Dublin, depicted a ghostly image of Christ standing in the midst of a Quaker Meeting. This Presence, I felt, was generally attributed the characteristics of a personal God with whom one could communicate directly and to whom one could direct one's petitions. It was not uncommon for the spoken Ministry at a Meeting for Worship to include a prayer such as the Lord's Prayer and sometimes a Friend might sing a hymn.

I enjoyed the fellowship and security of belonging to Friends, especially throughout my formative teenage years when the Young Friends group was central to many of my social activities. I also had the good fortune to attend a wonderful Friends Boarding School in Waterford. For me personally it seemed that, among the Christian denominations, Quakerism with its rejection of violence and its recognition of the value of all individuals was most consistent with the teachings of "real Christianity."

It was perhaps inevitable that the Quaker environment in which I was being nurtured would be tempered somewhat by the potent Catholic influence of some of my playmates, as well as by a zealous Catholic housekeeper who had charge of me as a young child while my parents were at work. At the age of nine or ten I recall being troubled by the sight of some of my friends blessing themselves on passing a Catholic church on the upper deck of a tram. Not wishing to appear to be an outsider, which, of course, I was in a way, and not feeling entitled to follow their example, I would pull my peaked school cap low over my eyes, scratch my forehead and fiddle vigorously with the buttons of my raincoat! I am not sure how convincing my performance was: my friends did not comment.

I settled quite comfortably into a belief system that reflected my early upbringing. As I progressed through high school and later, science studies at university, I did notice, however, that many of these beliefs did not seem to be in accord with common experience and did not, indeed, stand up to scientific scrutiny. I began to have difficulty with the concept of a personal God who interested himself in the affairs of humanity and intervened personally to control the course of events on Earth. I wondered about alleged miracles. I could, to some extent, come to terms with these difficulties by interpreting the phenomena in question as metaphors or symbols, but I was experiencing the first serious doubts about my faith.

The science program in which I was enrolled focused primarily on chemistry, physics and mathematics, all disciplines in which critical analytical thinking and reasoning are central. Progressing through the course, I found myself gravitating towards chemistry, especially organic chemistry and the chemistry of natural products.

One thing that impressed me early on was the way in which over the years methodology had been developed in the sciences that allowed the elucidation of the most intimate details about the three-dimensional arrangement of unseen atoms in a complex

Chapter 1

unseen molecule. Although unseen, the reality of the molecule is rigorously verifiable by the study of its behaviour; indeed it is on the basis of this behaviour that the molecular architecture can be assigned. Any behaviour or reactivity not consistent with the proposed structure would automatically cast doubt on its authenticity.

In the first half of the 20th century, the determination of molecular structure was very much based on "test-tube chemistry." The elucidation of complex structures without tools other than laboratory glassware, chemical reagents and the application of logic was a monumental scientific accomplishment by the early pioneers of chemistry. The range of reactions that a compound would undergo with judiciously selected reagents provided essential clues as to the structure. Further important evidence was derived from chemically cleaving the molecule into smaller identifiable fragments. Assembling all the pieces of the puzzle led to the complete molecular structure, which had to be consistent with all the evidence. The process often took many years of meticulous work and required the collaborative efforts of large teams of chemists. For example, unraveling the molecular structure of the ubiquitous plant pigment chlorophyll-a required more than 30 years of painstaking experimentation and detective work by numerous researchers.

In the course of my career as a chemist, the elucidation of molecular structure, previously often something of an intellectual tour de force, has become almost a routine process through the application of physical methods, including measurement of characteristic absorptions of electromagnetic energy and most notably by X-ray diffraction measurements. A structure that may once have taken years to elucidate can now often be solved in hours. It is a testimony to the genius of the early scientists that structures once painstakingly deduced on the basis of chemical reactions are now quickly verifiable by physical methods.

For me, the role of natural products in the coevolution of plants and herbivores and the study of chemical ecology have become fascinating and rewarding areas of research. Interest in these areas has grown sufficiently to spawn the publication of specialist journals. The study of natural products is, of course, just one of many branches of chemistry. The key fact for me is that the beautiful three-dimensional architecture of molecules, although they are unseen, does not have to be taken on faith but can be proved by subjecting the structures to the verifiable/falsifiable test.

Philosophers and theologians continue to debate whether such criteria of reality as testability/falsifiability can or should be used in an attempt to prove the existence of God. Some hold that the belief that the scientific method is the only reliable way of arriving at "truth" is itself impossible to prove.[1] A belief in the scientific method has often been pejoratively termed "scientism." These critics do not dispute the value of science as a profound and fruitful approach to learning about the material Universe: What they take issue with is the conviction that the scientific approach has a complete monopoly on access to all truth. According to their view, scientism is an unprovable belief system no less than religion. While recognizing this argument, I was, nevertheless, predisposed to the belief that it should be possible to apply scientific reasoning to religious questions. This intuitive feeling has had a major influence on my spiritual journey.

Early in my undergraduate career, I was unwise enough to participate in a debate in the venerable Philosophical Society (the "Phil") at Trinity College Dublin, in which the motion at issue was "That God does not exist." Because of my inchoate

Chapter 1

doubts about a theistic God, I agreed to be on the side arguing for the motion. My residual theism, however, interfered with my advancing a convincing case for the motion and because of this, as well as the fact that I had little experience in debating, I succeeded only in embarrassing myself and my team-mate. I was an easy target for the divinity students who were promoting the "for God" arguments. The opposing team, destined to join the clergy of the Church of Ireland (Anglican), won the debate handily. I did not take part in any more formal debates at university.

Eventually it became necessary for me to subject my ill-defined beliefs to a closer scrutiny. This occurred during a two-year sojourn in the United States as a research fellow, which happened to be at the height of the Vietnam War, when it was mandatory for all young men of my age to register for the draft. Naturally, I attempted to register as a conscientious objector. If my memory serves me correctly, the first question on the application form for CO status was "Do you believe in a Supreme Being?" My initial response was a qualified "Yes, I do," but as I perused the rest of the questions on the form I began to think about the draft board tribunal in front of whom I might have to defend my concept of a Supreme Being, as well as my other beliefs. Was my vaguely defined concept of a Supreme Being one that would be recognized by a draft board? While I certainly felt that I experienced "spiritual feelings" and emotions, I found that time and the scientific critical analytical approach that I had been taught to use had indeed eroded my faith in many of the tenets of mainstream Protestant Christianity.

I could not bring myself to accept the idea of a virgin birth or the resurrection of the body and I found myself devising rational scientifically acceptable explanations for many of the miracles. I found it difficult to comprehend how any Being could simultaneously listen and respond to the prayers of literally millions of people. Although I was an alumnus of Trinity College Dublin, no one had ever been able to explain to me satisfactorily what was the true nature of the Holy Trinity. I was finding it increasingly difficult to summon belief "to order."

I pondered over what the concept of God meant to me. I did not want to be disingenuous and I had an uneasy feeling that the draft board would not find that any Supreme Being other than a personal God fit their criteria. I asked myself what and why is God and again the question recurred: can one apply the same verifiable/falsifiable criteria to religious and scientific beliefs.

Religions have been a significant part of all cultures since the dawn of civilization, which attests to some common yearning or need in the human psyche. Among other rewards, which I shared, religions provide comfort and consolation for their adherents in a world that often seems hostile. In many cases, religions help to allay our fear of death and address our longing for immortality. The comfort and security of belonging to a coherent community is, of course, also a significant incentive to join a religious group. Perhaps most important, religious faith promises to help us find insights into the meaning of our existence. In the past, children raised in a religious family have usually tended to remain within the same faith group, ensuring its continuance and growth.

The sun was the most obvious object of worship for early civilizations, as they instinctively recognized its life-giving power and majesty. Then, inanimate objects, animals, and humans were also accorded supernatural status as local or universal deities and multiple deities were recognized. Some form of Earth Mother Deity has been important for many cultural groups.

Chapter 1

Religions have come and gone, though several have persisted for millennia and still claim large numbers of adherents. The God that is central to the Judeo-Christian canon is a supreme, intelligent, supernatural being, which is credited with having created the Universe and everything in it including, importantly, life. For the great majority of Christians, God is a supernatural personal God, one who takes an interest in his Creation and who intervenes in the affairs of humanity, who answers prayers, judges our conduct and rewards or punishes us accordingly.

In the Lord's Prayer, a petition is addressed to "Our Father, which [or who] art in Heaven." In early symbolic images, the Father is portrayed as an old man with a long flowing white beard, however few people, other than perhaps children, now see God in this way. After the sanctity of God's name and power to control future events are acknowledged, the prayer asks God to provide us with sustenance and to forgive us when we stray from the moral compass spelled out for us in the Gospels. As a quid pro quo, we undertake to forgive those who trespass against us. Millions of Christians add their own personal petitions to the Almighty, for instance for the health of themselves and their loved ones or for the success of various endeavours and, yes, for the defeat of their enemies. It is accepted that God can hear and respond to every one of these petitioners. The Lord's Prayer is a central part of the religious practice of most Christians. When I recited the familiar words, I asked myself who or what, in all honesty, was I addressing?

I was fortunate not to be called before the draft board and I finished my stay in the United States without a serious threat of being dispatched to Vietnam to fight in a war to which I was profoundly opposed. I had not found satisfactory answers but, one way or another, an important milestone in my spiritual journey had been reached, and I continued over the years from time to time to subject my ideas about religion and metaphysics to more serious scrutiny.

Another significant step on the journey, much later, was participation in an extramural course at the University of New Brunswick on science and religion. Many of the discussions during the course were stimulating, but some of the peripheral literature I found myself exploring was the real eye-opener for me – for example, *The Pagan Christ* by Tom Harpur.

Against the background of his belief that myth is more eternal in its meaning than history, Tom Harpur, a scholar of religion and former Anglican priest, argues that it has been proven beyond doubt that events described in the Gospels were, in fact, derived from ancient Egyptian mythology, originating millennia before the advent of Christ. A summary on the dust-jacket of Harpur's *The Pagan Christ* states: "Long before the advent of Jesus Christ, the Egyptians and other peoples believed in the coming of a messiah, a madonna and her child, a virgin birth, and the incarnation of the spirit in flesh. The early Christian Church accepted these ancient beliefs as the very tenets of Christianity, but disavowed their origin."[2] Harpur's book contains a quotation describing Northrop Frye's lectures in which he taught his students that "...when the Bible is historically accurate, it is only accidentally so: reporting was not of the slightest interest to its writers. They had a story to tell which could only be told by myth and metaphor." Harpur's revelations cannot help but cause one to view the Bible, with all its wisdom and poetry, from a new perspective.

And then there were the arguments of Richard Dawkins! I had been warned that Dawkins' attacks on religion were

Chapter 1

"vitriolic." Indeed, his book, *The God Delusion*[3] is an unapologetic, vigorous polemic against religion. It is in the nature of polemics to be uncompromising, even intolerant. There is no doubt that *The God Delusion* has caused offence to a great many people. Whereas one may be uncomfortable with the militant iconoclastic aspect of Dawkins' writing, he does present scientifically plausible explanations for the creation of the Universe and the development of higher life forms on at least one of its planets, without supernatural intervention. He also presents a sophisticated philosophical argument against the existence of a theistic God, which on the surface seems difficult to refute in logical terms.

The God Delusion covers much ground. Dawkins' summary of the origins of the Universe(s) as well as the origins of life on this, and probably other, planet(s), based on the statistics of astronomical numbers, as well as on the so-called anthropic principle, are credible, and seem to be in accord with the ideas of modern theoretical physics and cosmology. Nowhere in this account is it necessary to invoke the supernatural as a default explanation for any apparently improbable phenomenon.

Based on Darwinian evolution, mutation and natural selection, Dawkins' account of the derivation of higher life forms, including *Homo sapiens*, from single cell organisms is highly plausible, given the self-replicating properties of DNA and other macromolecules. It is, of course, possible to accept these momentous and crucial phenomena within a religious context, a fact patently not lost on Dawkins, but, as he argues, they are more likely to have occurred according to ongoing cosmological and biological evolution rather than Divine intervention.

Brian Swimme, a mathematical cosmologist, and Thomas Berry, a Catholic priest who calls himself a "geologian," lay out the same scientifically based narrative in their book, *The Universe*

Story, but with this difference: Rather than spending intellectual energy debunking the archaic notion of "Divine intervention," they see the unfolding of the Cosmos, planet Earth, and its great diversity of life as a single "revelatory" event that is marked in all its phases and expressions by a play of differentiation and communion.[4]

English playwright, Caryl Churchill, expresses this reality in a way that resonates with me and brings a smile. "We've got ninety-nine percent the same genes as any other person. We've got ninety percent the same as a chimpanzee. We've got thirty percent the same as a lettuce. Does that cheer you up at all? I love about the lettuce. It makes me feel I belong."[5]

Creationists stridently proclaim that Darwin was, in fact, wrong; that literal interpretation of the Bible gives a true account of the creation of Earth and all life on it. In their view, the complex life forms that are found on Earth could only have come about by "Intelligent Design," the Intelligent Designer being God. Dawkins shows that it is not difficult to utterly refute this view, which, nevertheless, continues to be held by a large number of fundamentalist Christian groups, especially in the United States. The majority of Christians nowadays, however, are comfortable with the idea that their God chose Darwinian Natural Selection as his modus operandi in the creation of the World.

But, it may be asked, if we do not need to invoke God for the creation of the Universe, the Earth and the life on this planet, why do we postulate the existence of God? I earlier reflected on some of the factors and perceived rewards that have contributed to the development and evolution of religions, including those of the Judeo-Christian tradition. None of these factors actually provides anything approaching an attempt to prove the existence of God.

Chapter 1

Personal spiritual experiences are often cited as proofs of God's existence, but skeptics will argue that even the most rational human mind is capable of conjuring imaginary phenomena under certain emotional influences. Undoubtedly, when George W. Bush looked the world in the eye and claimed that God told him to invade Iraq, it gave little credible support to the idea of personal "revelation." Nevertheless, personal spiritual experience is something familiar to all religious people and, while it may not "prove" the existence of God, it is too important to dismiss, certainly for Quakers, and I shall return to the topic later in this essay.

In his discussion of cosmology and evolution, Dawkins points out that God is invoked to explain the existence of things and phenomena that appear too complex to explain without supernatural intervention. If, he argues, these phenomena are too complex to arise spontaneously unless God intervenes, then God himself must be even more complex. How does one rationalize the spontaneous existence of a Being (God) who is even more complex than something that is itself considered too complex to arise spontaneously? Who created God? One enters an unsatisfactory situation of infinite regress.

Dawkins and other atheists conclude that the existence of God, though impossible to disprove conclusively, is highly improbable. The arguments seem convincing, in their way. Despite some of the more controversial aspects of his writings, a number of Dawkins' arguments have brought conclusions into sharp focus that I had gravitated towards via a less analytical and less confrontational route.

A personal, supernatural, theistic God no longer has a place in my belief system. This said, I share the widespread experience of feeling a sense of profound reverence and awe in the presence of great beauty in Nature or human endeavour. Thus, looking upwards at a starry sky or a spectacular cloud formation, enjoying the chatter and sparkle of a brook or the sibilant pounding of ocean surf on a beach, walking in the forest, exploring a backwater in my kayak, looking in a museum at an original Monet painting or listening to a great musical work by my favorite composers - Bach, Mozart and Beethoven - can all become "spiritual experiences" for me. As a young man, walking or climbing or simply being in the midst of hills or mountains had a special capacity to produce in me a feeling of being at peace and in harmony with the Universe. The Psalmist wrote "The mountains shall bring peace to the people, and the little hills, by righteousness." (Psalm 72:3 KJV). Although mountains in this context were perhaps intended as a metaphor for powerful rulers, the statement worked for me in a literal sense.

Many people would consider that such experiences are revelations of the workings of a personal theistic God. For me, they are experiences of being face to face with what I consider as "the Divine," that is to say, God and the Universe or Nature are indistinguishable. The term pantheism came to be used in the 17th century to describe such a philosophy, though the philosophy itself in various forms has been around since the pre-Christian era.[6] One sees some parallels in considering the "Earth Mother" as Deity. As a pantheist one is in good company. Great thinkers, poets, and scientists, including Spinoza, Walt Whitman, Ralph Waldo Emerson, Henry David Thoreau, and Albert Einstein, to name but a few, espoused or contributed to versions of pantheism.

Einstein, "a deeply religious nonbeliever,"[7] expresses this kind of experience with clarity and humility: "I believe in Spinoza's God who reveals himself in the orderly harmony of what exists,

Chapter 1

not in a God who concerns himself with the fates and actions of human beings." ... "We followers of Spinoza see our God in the wonderful order and lawfulness of all that exists and in its soul as it reveals itself in man and animal."[8] Elsewhere he writes: "I don't try to imagine a personal God; it suffices to stand in awe at the structure of the world, insofar as it allows our inadequate senses to appreciate it."[9]

The Universal Pantheist Society was founded in 1975 and formally used the term "religion" to describe its beliefs. The World Pantheist Movement was created in 1999. The beliefs and values of the latter group aim to "reconcile spirituality and rationality, emotion and values, and environmental concern, with science and respect for evidence." These beliefs and values are summarized in a Statement of Principles:[10]

- Reverence, awe, wonder, and a feeling of belonging to Nature and the wider Universe.

- Respect and active care for the rights of all humans and other living beings.

- Celebration of our lives in our bodies on this beautiful earth as a joy and a privilege.

- Strong naturalism - *without belief in supernatural realms, afterlives, beings or forces.* [my italics]

- Respect for reason, evidence, and the scientific method as our best ways of understanding nature and the Cosmos.

- Promotion of religious tolerance, freedom of religion, and complete separation of state and religion.

Dawkins,[11] Weinberg,[12] and other avowed atheists have no quarrel with the philosophy of pantheists, but they argue that such beliefs do not fall under the conventional definition of religion, that is to say theistic religion, involving belief in a supernatural god, that they seek to debunk. Recognizing the agenda that underlies their definition, I would argue for a broader, more inclusive definition of religion: one that would encompass any coherent shared belief system or philosophy that helps to provide profound insight into our role in the Cosmos and a moral compass to guide our lives.

Whether one is atheist, secular humanist, pantheist, or theist, many universal questions remain a mystery. Our increasing scientific knowledge about the neurochemical and neurophysiological mechanisms associated with the workings of the brain, as well as the parallels we can see with computers and artificial intelligence, seem destined to give us a rather complete understanding of this magnificent organ. The concepts of mind, consciousness, awareness, and spirit straddle the biological disciplines as well as psychology, philosophy, and metaphysics. Even if all the mechanistic aspects of brain function can be elucidated, the true nature of consciousness and awareness remains elusive.[13]

What is the source of the thoughts that are processed inside our heads? What is thought? Where does creativity come from? What is the origin and nature of what is widely accepted as "spiritual energy"? Not wishing to revert to a kind of "god of the gaps" belief,[14] which, by definition, diminishes as our knowledge and understanding increase, I would simply say there are a multitude of things that we do not, at present, understand and I suggest that there are some that may even remain for ever beyond human understanding.

Chapter 1

It seems to me there is an almost whimsical irony in the fact that many of the ideas of modern theoretical physics and cosmology about matter and energy, such as string theory and multiverses, may turn out to be as difficult to prove using the traditional scientific approach (verifiable/falsifiable) as proving the existence or non-existence of God.

I feel very comfortable with all of the principles espoused by the modern pantheist movement (see above), but I have no formal affiliation with either the Universal Pantheist Society or the World Pantheist Movement. It is my experience that one who may lean strongly towards pantheist philosophy can find their spiritual home in the Society of Friends. Many other members of the Religious Society of Friends have travelled spiritual journeys which have also ultimately converged on an essentially pantheist worldview, notable amongst them the American Quaker writer and naturalist Sharman Apt Russell.

In a memoir entitled *Standing in the Light: My Life as a Pantheist*,[15] Russell provides a brief and informative history of the development and evolution of pantheist philosophy from its origins in early Greek Epicureanism and Stoicism through the ideas of nineteenth century and contemporary pantheists. Not surprisingly she devotes a chapter to the seventeenth century philosopher Baruch Spinoza who was a pivotal figure in the development of pantheism and was excommunicated for his ideas by his Jewish community. Spinoza is known to have interacted with Quakers and to have felt an affinity with Quakerism.

Sharman Russell's historical and philosophical elucidation of pantheism is augmented and leavened by being interwoven with lyrically written personal reflections of life in the mountains and deserts of southwestern New Mexico and the natural history of the region. This background provides the essential nourishment for her spiritual journey. I identified strongly with her sense of awe and connectedness with the universe.

For further reading I also commend the book, *Godless for God's Sake, Nontheism in Contemporary Quakerism*, in which twenty-seven Quakers from thirteen Yearly Meetings in four countries describe the spiritual searches that have led them in various ways to affirm a deep and meaningful understanding of religious life that is not tied to the traditional belief in a supernatural, transcendent, personal God.[16]

For me, it is sufficient to know that deeper levels of consciousness and awareness are accessible through meditation, either alone or in fellowship with others. If we can "tune in" and be in harmony with our inner selves, with the people around us, and with the universe of which we are an infinitesimal but real part, all of our spiritual needs should be satisfied.

Imagination and Belief
A Composer's Journey Through Religion and Science

Michael R. Miller

Introduction

Just as we were driving away from her house, my sister-in law offered me an old paperback through the car window. It was *The Power and the Glory*, Graham Greene's 1940 novel.[1] She had begun to sift through her possessions, disposing of things she couldn't use again. It was lucky for me I accepted her gift. In my younger days I heard of Greene as an acclaimed author, but I never got around to reading what is considered his best novel. After arriving home I couldn't put the book down; its story and manner of telling were that intriguing to me.

I should admit a strong personal interest in how the author portrays the faith of a Catholic priest. Greene converted to Catholicism at about the same age I left the Catholic Church, he in the 1920s, and I, thirty years later. Apart from being a good read, this novel throws into high relief the contrast between the Catholic faith of the anti-hero and my position as a freethinking Quaker. My loving Catholic parents were quite relieved when I joined the Religious Society of Friends in 1975, along with my wife Edith. They were just glad we had become "something."

Chapter 2

The Power and the Glory recounts the last months in the life of a fictional character, a Roman Catholic priest, during the violent, anti-religious revolution in Mexico in 1938. It shows how this poor man was hounded from place to place by soldiers and police intent on executing him because he refused to renounce his vocation. At the same time he is conscience-stricken for becoming a "whiskey priest," addicted to brandy, living in sin with a woman, and fathering an illegitimate child whom he loved but could not support.

The priest, despite his human failings, adheres strictly to Catholic beliefs. He believes he was given specific mystical powers when he was ordained a priest. They include: changing bread and wine into the body and blood of Christ during Mass, the power to cleanse infants from Original Sin through Baptism so their souls won't go to Limbo, and granting death-bed absolution of mortal sins (as a last-minute escape from Hell). He also feared he was doomed to Hell because he was not sorry for siring a daughter.

Greene shows how much Catholic belief is centred in the supernatural and in the afterlife. Far be it from me to denigrate such beliefs, so long as they do not cause harm or inhibit factual truth. Am I a less-worthy human being because I don't make the countless assumptions, stated and unstated, that a good Catholic does? I don't think so. And I don't feel guilty about being disloyal to the religion of my birth, and for changing to another one in midlife. Although children need moral guidance and a sense of belonging as they grow up, they should be allowed to reassess their birth religion and to discard it or choose another religion or ideology upon reaching maturity. Of course, there is another option: to modify one's birth religion to accord more with new insights and knowledge acquired as an adult.

Is "Faith of Our Fathers"[2] Enough?

Is faith of our fathers enough? The quick answer, I would suggest, is - no. It has become increasingly clear that the traditional conceptions of the Jewish, Christian, and Moslem God are no longer adequate for the world of the 21st century. Frankly, these beliefs cause as many problems as they help solve. Am I saying that God as a belief or a fact is no longer tenable? Again, no; but our ideas of God and our assumptions regarding the Divine need to change radically if we are to have a theology worth believing and a future worth living.

Need for Examining Our Beliefs

Why am I getting mixed up in this God stuff? I am not particularly qualified in theology or philosophy to delve into this topic. I do have a Ph.D., but it is in musical composition! Writing down musical thoughts comes more easily to me. In fact, I would much rather be doing what comes more naturally to me – composing music. Why can't I leave these questions of metaphysics, theology, and philosophy to the meta-physicists, theologians, and philosophers and just do my own thing – play the piano and keep composing lots of music? But there are more sides to me than being a musician. I thirst for the widest and deepest meaning life can give me. I'm sure I'm not the only thirsty one.

As a youth, I was very lucky; my parents supported my musical interests. My father bought me a reconditioned Steinway baby grand when I proved serious about music. My mother thought that artists could understand God better because they shared

Chapter 2

the creative impulse. Perhaps a deep connection between artistic imagination and religious feeling and thinking can be traced to parental guidance. I have been jotting down various and changing thoughts on all this for many years. It's time for me to put things in order.

Metaphysics is the search for the eternal, universal nature of things. We can't talk about the Divine without making metaphysical assumptions. With so many competing claims for truth and value in today's world, it is important to think through metaphysical questions at a deeper level. Otherwise, some preacher, priest, or politician can mislead us. Simply ignoring these questions, as many people do, is being irresponsible to society, and especially to youth.

At the heart of the problem is the cloudy relationship between God and us. At this point in human history it can be characterized as a power struggle similar, I imagine, to the stormy relationship between parent and adolescent. In previous times of faith it was more like the life-supporting relationship between a little child and its parent in which parental wisdom, power, and goodness were not questioned. Could the parent-offspring comparison be applied further? We all know that it is possible, though not inevitable, that as the child matures and the parent mellows, a beautiful relationship of mutual respect and appreciation may grow. Dare we apply this to God?

I can already hear many objections to this parent-child analogy from both the scientifically minded and the religiously inclined. Some would probably say, "This is the kind of bunk you would expect from those undisciplined artsy-fartsy types." The critics would mostly come from two camps: those who would say that the whole God question is unimportant today and those who would maintain that God is the most important issue but that the "faith of our fathers" is what counts. Both of these extreme positions show a narrow-mindedness that does little to further an understanding of an issue of such relevance today.

My simple analogy of a maturing relationship between offspring and parent may or may not be helpful. But I cannot help feeling that we "moderns" are losing a great opportunity to further our humanity by not freely sharing our experiences of faith and doubt in God. Of course this goes against political correctness and may not always be prudent. Yet in the long run, wouldn't the benefits outweigh the risks?

Raising doubts about one's religious beliefs can be disturbing to some people. Religious assumptions are often embedded in one's ethnic or national identity. But inevitably the pressures of the new secular-global-consumer society must be faced one way or another. To retreat from society and live in a small, isolated group is not the answer for most people. Allowing one's attitudes and behaviour to be dictated by popular pundits or by the entertainment industry is not a good option either. Even if we find the guidance of counselors or religious leaders helpful, we are called on more and more to think and feel through questions of belief and values for ourselves.

When it comes to religion, many adults still seem to think like children, though in other phases of their lives they are quite mature, thinking things through for themselves and making their own decisions. Some people may feel that questioning religious beliefs offends God or is a sign of disloyalty to one's birth religion, country or culture. It may take a degree of patience and courage to find deeper ways of viewing the world, and of finding positive and responsive ways of living in it.

Chapter 2

Not Just for Experts

Could it be that my subject is of the very kind that should not be left entirely to experts? So, calling on other non-professionals to join the fray, I will bluster on, undeterred. If my thoughts seem a little strange or unhelpful to some, at least I will have satisfied a long-held urge to sort out my many notions by getting them on paper. And I will have "faced the music" in my own way. Of course the thoughts of others have been a great help. Of special note in this regard has been Chet Raymo's book, *Skeptics and Believers: The Exhilarating Connection Between Science and Religion* (1998).[3]

What do I mean by "facing the music"? Two things: 1) being open to the expanding insights and perspectives that science offers relating to our human nature and to Nature herself including the universe, and 2) trying to integrate these into a common world view in a way that enhances our humanity, that develops what is most precious in us individually and collectively. Putting it bluntly, this simply means that if we really want a happy, peaceful, just society to develop, we can no longer use religion or science as an excuse to kill, hurt, suppress, or exploit human beings.

Humans, the Earth, and the Universe

Every part of the Universe is formed of nothing but matter and energy including me, thee, and the All, or God. Such is one of my beliefs now. Before going on, let me back up at bit. It all began with the Big Bang about 15 billion years ago. At first the Universe was unimaginably compressed, hot and dense; ever since it has continued to expand, to cool, and to become more diffuse. Such a process will continue until the Universe dissipates all its energy and dies in a Big Whimper billions of years hence. Or at some point, due to causes yet unknown, the process may reverse itself, leading eventually to another Big Bang and another cycle of consequences.

In this science-based scenario, as contrasted with mankind's previous creation stories, humans no longer occupy centre stage in the cosmic drama. Instead we seem to have been assigned a tiny bit part that spans only a fraction of a second – so to speak. It is not only Genesis, the first book of the Judeo-Christian Bible that now appears incredibly anthropocentric, but also Revelations, the last book.

This is one of the more obvious contradictions between modern scientific thinking and traditional religious teaching. Less obvious perhaps is the difference in how we understand a religious teaching and a scientific fact. Usually we take religious teaching on faith and out of respect for authority and tradition. Behind scientific fact is rigorous and repeatable testing or at least a calculation of probability. Religious teaching often claims absolute and eternal validity; good science never does.

Here's another fanciful analogy: Compared to other animals, humans are Nature's orphans! We appear to have been abandoned and denied the guidance that developing children need. Instead we have been kicked out of the natural world and forced to make our own way. No wonder we make so many mistakes, more mistakes it seems than our non-human brothers and sisters, who simply accept their roles of struggle and cooperation without asking why. On the other hand, the success of humans as a species has allowed us to exploit every corner of the globe. We have thus also become Nature's Spoiled Brats!

Chapter 2

Science has shown us that every creature has made unique adaptations to its environment in its competition to survive and reproduce. Humanity's unique adaptation is imaginative intelligence. Armed with this, humans have won the competition against other beings, hands down. We are clearly on top, at least for now. But our success has come at a price. With godlike powers have come godlike responsibilities. We have still to find our true place in the natural scheme of things and act accordingly. Up to now with few natural impediments except for diseases and natural disasters, we have found it hard not to abuse our powers. But now, we have arrived at a time when we must stop expanding our world economy and our total population. If we don't, our future is bleak.

One does not have to look far to discover that many people are looking for guidance and clarity on theological questions and often find unsatisfactory answers. There is a general reluctance to discuss religion. It is considered a personal matter, quite likely to create division, even rancour between people. My feeling, however, is that there is a global need to get to the root of this human desire for the Divine. I dare say an inflexible refusal to examine one's beliefs contributes to inter-cultural conflict and even violence in the world today.

By now readers will probably assume this essay is headed toward the customary defining, contrasting, and polarizing of religion and science. It is easy to side with one or the other, claim that never the twain shall meet, and end the discussion. However, I cannot help believing that much good could result by seeking and imagining ways in which each worldview could enlighten the other. Here are some suggestions that might help:

- Both science and religion could be less arrogant in claiming exclusive access to the truth. They could both admit their mistakes and abuses with more openness and less cover-up.

- Religion could become less authoritarian and accept impartial evaluation more willingly.

- Science could acknowledge that individuals and societies can benefit greatly by shared ethical beliefs that cannot be proved scientifically.

- People with either bias could take themselves less seriously and learn to laugh more - even at their own expense.

My aim here is to share my reflections and inventions with others. Even though they are personal, my hope is they may strike a chord (or even a discord) with others and provoke a response. We learn as much being shown wrong as right. So, here in Friendship, I jump right into the middle of where "angels fear to tread."

Does God Create Us, or Vice Versa?

Does God create us, or vice versa? My answer to this question is – a bit of both: God creates us and we create "him." By saying that God creates us I mean that individually and as a species we are formed by our evolutionary inheritance. When I say, "we create God," I mean our concept of God is formed by our cultural and religious inheritance. As humans evolved into questioning beings, it was natural that they thought supernatural beings were behind all the mysterious, wonderful, and terrifying phenomena of life.

Chapter 2

Unlike other creatures, which seem to know and accept their roles in the world, humans have had to be more inventive. Only humans find it necessary to worship gods. Why are we so lacking in self-confidence that we crave the blessing of some super being? What makes us so dissatisfied with this life that requires us to believe in an afterlife? The short answer is the human brain allows us to do so.

In the view of science, and widely accepted by many learned people, we were not "created" in the biblical sense, but came about through evolution in a long and complex manner. However, the process of natural selection implicit in evolution no longer operates in the case of humans: We can survive and reproduce despite the disadvantage of traits and conditions that would not allow survival in the natural world. The human species now is, as previously mentioned, a kind of orphan, abandoned by Mother Nature. What I mean is that humans can no longer expect instinct to be their only guide. This leaves individuals and societies with an increasing responsibility for moral choice and action. We have all seen in our lifetimes the power that science and technology has recently given to humans to wreak harm on their fellows and on the planet. It is easy to lose sight of the good this new knowledge and power has done and can do in the future if properly and justly used.

I have a definite preference for the evolution story over the creation story as an explanation of how the Universe came to be, including humanity as an unimaginably small part of the total. I expand the term evolution to include the universal process of development that began with the Big Bang and continues toward an unknown future. It is defined by much transformation from the simplest constituents of inanimate matter/energy to the most varied and complex life forms.

Imagination and Belief

By the creation story I mean more or less the belief – traditional in most cultures – that humans, animals, and plants were made by a super being or beings that resemble humans but live longer. "Intelligent design" is the assumption, probably dating back to the scholastic philosophy of the Middle Ages, that a supernatural being must have created everything including humans because only this explanation seems possible. Who or what created this supernatural being is a question left without an explanation.

In my youth as a Catholic I remember being told by priests that as long as one believed that the Universe had been created and continued to be run by an all-powerful, eternal, pure spirit whom English speakers call God, one was at liberty to believe He used evolution as a kind of tool to form animate life. How the inanimate part of the Universe was formed only God knows. This was before the Catholic Church had tried to assimilate nuclear structure and the Big Bang into its official teaching.

The question, "does God create us or vice versa?" is a paradox. My answer, "a bit of both," is also a paradox – a "most ingenious paradox," as Gilbert and Sullivan would sing. *The American Heritage Dictionary of The English Language* defines "paradox" as "a seemingly contradictory statement that may nonetheless be true."[4] For me, a paradox is the most appropriate response for what is total mystery, a mystery summed up in the three questions of the title of Gauguin's famous painting: "Who are we? Why are we here? Where are we going?"

Why not just let the mystery be and refuse to define or even express it? Because a mystery to the human mind is like a vacuum to nature: It yearns to be filled. Freedom and consciousness are characteristic human qualities, but these have terrifying implications if regarded as absolutes. Like everything else in the Universe humans are evolving, hence unfinished and imperfect as is God himself. Hence his forgiveness; he forgives himself as he forgives us while he asks us to forgive him. Forgive him for what? For foisting this load on us, this responsibility for consciousness, for giving us a part in this creation play. Who wouldn't get stage fright?

God creates us in the sense that practically everything we are or have has been given to us - our parents, our genes, our psychological make-up, our social standing, and, yes, our good and bad fortune. How we think and act is affected but not determined by these givens. What we can control the most are our attitudes. Although humans do act stupidly and contend with self-pity, hatred, fear of the future, envy, and other negatives, these unfortunate attitudes imply positive qualities that are also typically human - a feeling of self-worth, altruism, optimism, generosity, and other life-enhancing attitudes. To me, our attitude toward life is an area where we can all be creative artists, heroes, and saints. We are called by destiny to continue this work-in-progress all our lives. Its subject is our interaction with our fellows, our culture, and our ever-expanding world.

Regarding the "vice versa" of creation, I did not intend any pun to indicate I disapprove of creating God, as sacrilegious as it might seem to some. Perhaps, we can say with Meister Eckhart, the 13th century Dominican mystic, "The eye with which I see God is the same eye with which Gods sees me."[5] I believe that God is a concept that humans have invented to answer the mystery of our existence. My sense is that what is at work here is a deep biological awareness found in the simplest life forms that is also present in human consciousness. It is behind such human attributes as curiosity and common sense.

Let me explain: Even an amoeba knows the boundaries of its tiny body.[6] All living things would disintegrate if they did not possess this basic self-awareness. This of course implies an equal awareness of the world outside of its body - food or poison, friend or foe, vital to its survival. Thus from the beginning, life divides the world in two: the other and me. It is thus very difficult, indeed against common sense, to conceive of ourselves as part of the other, and the other as part of us. But if we stand back and try to view the me and the other from outside, we may see that both are part of a system, dependent on each other, like the components of a natural environment. Is it so unreasonable to suppose that "creation" and "creator" is one system?

We create God in original and unoriginal ways. The unoriginal way is when we accept as children or converts what we have been taught by our parents or teachers. Originality starts to enter our creation of God as our experience of life, and our deepening knowledge of the world prompts us to test the validity of received beliefs. We become skeptics. Skepticism can be constructive or destructive. Constructive skepticism helps create new authentic religious belief that imparts the great psychic energy to individuals and societies, allowing them to overcome great difficulties and to build wonderful cultures.

Imagination - Behind Culture and Religion

One of the powers that intelligence brings is, of course, imagination. An example of this is our almost universal traditional belief in a god or gods. I have a hunch that it is our separation from the natural world that drives us to imagine immaterial beings that control the natural world. Fears of death and catastrophe and the desire for reassurance against these fears prompt a belief in the supernatural. This belief tends to quell the apprehension that the supernatural may originate in our imagination.

When I think of religion in connection with imagination, the Old Testament stricture against making "graven images" comes to mind. To me, words, ideas, customs, and attitudes are images whether or not they are depicted in a statue or picture. It is wrong to forget that we create them, wrong to begin worshipping them as something divine and unchangeable. When such images are no longer sufficient or relevant, they should be replaced or allowed to evolve in keeping with the new reality. This is not to say that they should not be preserved for their historical or artistic value.

"Image," naturally, is the root word of imagination. It is interesting to note that the word "religion," literally and simply means, "to tie together."[7] Perhaps an implied meaning is to give form to one's impressions, to make sense out of experience. For imagination to work, it has to have been "hard-wired" into our brains and nervous systems. Could imagination be a by-product of the ancient shrewdness we needed to survive against bigger and stronger species like the mammoth and sabre-toothed cat? Imagination helps us create language, culture, religion, government, commerce, technology, and science. It allows us to be the supremely social and successful species we have become.

When we think about God, we create a mental image of Him/Her/It whether we realize it or not. Even our factual knowledge of the world as discerned by our five senses is to some extent imagined. Abstract concepts such as good, evil, and God are imagined to a greater extent. As long as we remember that our human gift of imagination is operating here, we can accept more easily other, even contradictory, views of God. In my view, religion has been far too slow in accepting the tremendously expanded mental and spiritual horizon that science and cultural history have opened up for us.

What I am about to propose may appear to be of doubtful value for some. Indeed, some people may describe these ideas as atheist, pagan, animist, heretical, and non-Christian. Yes, the following is an image of God that is in conflict with many traditional assumptions. I make no claim of originality, though I have not heard or read of these ideas held by others. Actually, it would be better if they were not too original, as they might then be more readily accepted. In any case, my contention is that we are in dire need of a new theology, one that accords better with the experience, needs, and knowledge of human beings living in the 21st century.

Here is how I wish we would imagine God:

- God is the All of reality – what we know and understand and what we do not.

- God has a spirit and a body. The spirit of God is the energy of all matter, and the body of God is all the matter of the Universe, inanimate and animate. Spirit and body, like energy and matter, are two aspects of the same entity.

- Perhaps the Big Bang (15 billion years ago) marks the point at which God, who exists as eternal possibility, became real by taking on matter and energy, and entering space/time. From this moment on, God Itself evolved from the tiniest space possible to the ends of the Universe. God began as the simplest sub-atomic particles, evolved through atoms, star dust,

Chapter 2

stars, planets, molecules, one-celled life-forms, plants, animals (including humans), and complex social forms that transcend individual organisms. (This account, of course, leaves out many stages and branches.)

- Chance, however, plays its part from time to time. Though it may temporarily halt development, it seems also to stimulate further development. The four great extinctions of prehistoric life are examples of this. God's passion for development seems to win out in the end. He also likes to remain simple: the most common form of matter/energy is hydrogen, the simplest atom, accounting for 99% of the material in the Universe.

- Nothing is formed *ex nihilo*. Simple forms are combined into more complex forms. Complex forms constantly fall back into simpler forms. For example, this happens in the animate world at death, and in the inanimate world when molecules break down into smaller molecules or atoms.

- God is subject to the negative aspects of existence: death, imperfection, diminution, destruction, decay, and failure. These negatives, including death, are partial because the simplest elements are never destroyed and have existed since the Big Bang.

- God as matter/energy had a very violent and chaotic beginning, but as time unfolded became more orderly, transforming into patterns and more complex forms. In the words of Thomas Berry, "The Universe has a violent as well as a harmonious aspect, but it is consistently creative in the arc of its development."[8]

What are we to conclude from this view of God? That to exist in reality and to evolve is God's destiny as it is ours as human beings. Whether God materialized by accident or design is not important. We must simply realize that as imaginative beings on a tiny planet in an obscure solar system in one of billions of huge galaxies, we are challenged by being - as far as we can tell – one of the most developed inheritors on Earth of God's evolution. What a responsibility!

When I describe humans as the inheritors of God's evolution, I mean that we have, like other beings, our own characteristic way of participating in the unfolding of the Universe. As we look around us in the natural world, we see that all living creatures great and small have a fierce desire to live, to prosper, and to reproduce their kind. This occurs despite the suffering and death that accompany life. Even the inanimate world shows a remarkable tendency not only to exist but also to become more complex, more interesting, more varied, and more artistic.

Humans, of course, have the same strong desire for life as other conscious beings, but through evolution we have also acquired the ability to be self-conscious. We can stand apart from ourselves and imagine our situation in the world as different from reality. Especially as we mature into adulthood we can reflect on our human situation as Hamlet does in Shakespeare's famous play when he says, "To be or not to be, that is the question." In choosing a life of engagement with the world over one of despair and self-destruction we can also choose being over not being. Moral choices, like other choices, are possible in humans because of our intrinsic imaginative ability. I believe that imagination is behind everything that makes us human, not just the arts. Like other abilities it may be used well or badly.

Chapter 2

Critique of "Old Time Religion"

I no longer believe as conventional Christians do, that there is a world beyond the physical planet Earth and the Universe. I do not believe that we survive death as individual persons. I believe that living this life as consciously and responsibly as I can with the invaluable help of others is better than hanging around as a ghost forever after (whatever that means). I can't help thinking that a preoccupation with a personal afterlife shows that our ancient instinct of self-preservation can still overpower our more recently evolved rational sense. It is not surprising that our ego and our fear for its demise should cause us to invent a world of wish fulfillment in which our ideas of justice and self-importance would be perfectly corrected.

The Christian heaven is a kind of second Garden of Eden that blooms forever somewhere beyond the clouds, but is just big enough for those who have been "saved" by Christ. This story implies that humans deserve a perfect world of peace and plenty, where men do not have to work for a living and women do not have to bear children in pain. It seems as if the Old Testament writers did not realize how unreasonable was this expectation of such a utopia where every legitimate human desire and need is granted and every pain and fear banished. Talk about unconscious egotism, personal and social! What an example of man's anthropocentric tendency! While such myths hardly reflect reality, they certainly show us much about ourselves.

The Eden story also tells us a few other unpleasant things about ourselves. It shows that when we don't get what we expect, we look for someone to blame and punish. The prime suspect, I would expect, would be God who, fully aware of the curiosity and enterprise of humans, emphatically forbade "our first parents" to

Imagination and Belief

eat the fruit of the tree of the knowledge of good and evil. As if a desire for knowledge were a sin! But we all know that poor, naive Adam and Eve are blamed for thinking for themselves and punished by having to experience pain and death like the rest of creation and by banishment from a perfect utopia.

What is even harder to accept is why God put his curses on all the descendants of "our first parents"? I guess having perfect alibis doesn't count in God's court. Not exactly a picture of a just God! Of course the creators of this tall tale and its believers reflect the family, social and political assumptions of their times and places, ancient and modern. Such people are used to accepting the authority and power of their paterfamilias or sovereign without question no matter how unjust or arbitrary.

It is interesting to note that in Christian theology the one, king-like God becomes more like a caring father, and acquires two more persons: the Son and the Holy Spirit. Most importantly, God becomes human as in some pre-Christian mythologies, and, as Jesus, ministers to the people of his time and place as recorded in the New Testament. Despite this evolution in the positive image of God, certain ancient ideas persist in the Christian world view: 1) that the Divine essence is pure spirit, though It will manifest Itself physically from time to time in Its relation with humans, and 2) that human sacrifice is necessary to appease the Divine for offenses against It. I can't help thinking that such beliefs are unproductive to say the least, and positively dangerous when taken seriously.

And finally, a big mistake that we humans have made since time immemorial, and that Christianity has perpetuated, is to assume that the spirit that appears to animate the world and its creatures is "separate and other" than the world and us. In my view, God does not exist at all if She/He/It does not have a body. God's body is all the matter in existence, and all the energy that exists is God's spirit. The Universe as God's body unfolds itself in an orderly way that makes sense, whether we can discern it or not. It is wonderful but not miraculous or arbitrary. It is not perfect and unchanging, but it does follow certain laws. It seems to be eternal and is in constant change. If it did not change it would not exist! To me, Creator and Creation are one and the same.

The Human Jesus

What about Jesus? He was a great human being with extraordinary and revolutionary ethical teachings for his time. He spoke out against prejudice, injustice, and hypocrisy. But Jesus did not tackle such evils as slavery, war, or the inferior social status of women and children.

I am much attracted by his message of love, forgiveness, and caring for others. Many of his parables I revere. I first started appreciating these stories that stimulate the conscience as an adolescent at Sunday Mass in my Catholic days. I wish they were better known today by all children, not as religion but as part of cultural heritage.

Orthodox Christianity accepts Jesus as a saviour, a saviour from the guilty, fallen state caused by Original Sin. In order to give mankind a second chance at getting to Heaven, Jesus as God's son takes on the punishment for the sinfulness of human beings by dying on the cross! I interpret this dogma more liberally by basing it on older stories about the death and resurrection of pagan gods. Perhaps such tales are telling us that the Divine is in Nature and in us as part of Nature. As such, God suffers and dies and keeps on going, evolves and devolves and survives like everything else in some form or other.

However, one can also regard Jesus as a compassionate teacher of ethics, preaching forgiveness, mercy, and kindness. For example, he urges anyone with a grudge against a neighbour to become reconciled before making an offering to God. In the Beatitudes he says, "Blessed are the merciful for they shall obtain mercy." One famous parable is about a man who is attacked by robbers, who leave him at the roadside, stripped and beaten. A Samaritan, considered inferior by society, rescues and restores the victim after two respectable men had passed him by. The kindness of the Good Samaritan is contrasted with the cold-heartedness of the others.

In seeking followers Jesus casts his net wide enough to include such undesirables as tax collectors, sinners, thieves, and prostitutes. He even values the faith of a foreigner, a Roman soldier above that of "the sons of the kingdom." Hard to miss in the Gospels is the bias toward the poor lower classes and against such elites as the scribes and Pharisees, whom he terms a "generation of vipers."[9] The whole chapter of Matthew 23 recounts his condemnation of them for their cruelty, hypocrisy, and abuse of power.

Jesus was very much a man of his time and culture. He certainly had a sense of mission. He was leading a moral and political revolution. It was to overthrow and punish a decadent theocracy and to replace it with the "Kingdom of God," a kind of Eden, another perfect world. Furthermore, he promised this would happen before the demise of his generation. Obviously, this divine revolution did not occur. Somehow the later followers of Jesus did not take his word "generation" literally and thought he was referring to a Second Coming. To do otherwise would have called into question Jesus' prophetic powers. The Second Coming is supposed to occur at a future date. It is another perfect world where the good guys win and the bad guys lose. Alas, the desire for perfect worlds runs deep in the human heart.

Perhaps we are now beginning to understand how viewing the world in such black-and-white terms can do more harm than good. For example, such simplistic beliefs can inspire young fanatics to commit murder and mayhem in the name of religion. These beliefs can also keep us from making practical improvements in our lives by ourselves as individuals and as societies. For instance, if we count so much on God setting everything to right in the next world, why should we bother trying to improve this world?

The actual facts of Jesus' life will never be certain despite all the historical research and Biblical scholarship of the last 75 years or so. According to the stories about him, Jesus did not hesitate to work "miracles" in order to get the poor and powerless to listen to his message of hope and forgiveness. His followers seemed to come mostly from the working class. It is hard to avoid the impression that Jesus thought of himself as the Messiah and of his mission as the beginning of a new era. Apparently he was convinced that his mission was to improve Israel morally and to establish a sort of utopia: "My kingdom is not of this world."[10]

The writers of the New Testament saw no need for an objective account of a person so obviously worthy as Jesus, nor of the events of his life and work, which were so demonstrably good. They had a different conception of truth than the modern one. To them what they considered wise and good and what they considered true were indistinguishable. They did not value the kind of dispassionate objectivity that modern science has shown

Imagination and Belief

to be so valuable. In short, the ancients considered history and story to be practically interchangeable.

New Testament scholars tell us that people of the time, both Jew and Gentile, believed literally in miracles, i.e., astonishing suspensions of reality, magical happenings as found in folk and fairy tales. While Jesus is recorded as performing miracles, he is also recorded as rebuking those who demanded "signs and wonders" from him, rather than having confidence in his wisdom and goodness as such.

I find it hard to understand why Jesus accepted his crucifixion and death so passively, even fatalistically. Why did he not defend himself at his trial? Why did he think God wanted him to suffer and die rather than live and continue spreading his liberating message? Would it not have been more responsible of him to stay around a little longer if only to keep an eye on his less-than-brilliant and unreliable disciples?

A Catholic priest once said to me that we had to regard Jesus as divine or "crazy!" I disagree. Why not just accept him as human. The Quakers say there is that of God in every person. I would add that in the case of Jesus, there is more than the usual amount of God, "crazy" or not.

What About Morals and Ethical Behaviour?

What is the connection between beliefs, morality and ethical behaviour? Is this what 18th century philosopher Voltaire was thinking about when he said, "If God did not exist, He would have to be invented."[11] The more I have struggled to express my

Chapter 2

deeper thoughts and feelings the more I have come to see the wisdom of that remark.

There is a subtle, though essential, connection between ethics and God or the Divine. At this point, for the sake of clarity, I would simplify God as the Other, and ethics as how we treat others. I realize that God or the Divine also includes ourselves, and how we treat ourselves. But our natural instinct of survival tends to support our personal and group interests anyway, while belittling the interests of those outside our group. As a sense of the Divine, the Other serves as a counterbalance to this "natural" egotism and helps us remember to be considerate of others, which includes everything from our threatened environment to our threatened fellow humans. I think Jesus implied this when he said, "He that loveth not his brother whom he has seen, how can he love God whom he hath not seen?"[12]

To me, the place to begin is with values rather than beliefs. Values are obviously more social in character and have more to do with ethics. By starting with values I am not implying that beliefs are less important than ethics. Beliefs seem to be connected with the deeper layers of the psyche and reflect the ways in which we assume we relate to nature, to life and to death. In a healthy multi-cultural society a consensus on beliefs is not necessary, but agreement on values and certain ethical principles is. In looking for consensus in our diversity it is simply more practical to begin with values.

Here are seven of my most important values:

- Forgiveness of others and ourselves for acts that range from the mildly annoying to the horrific: This is not to imply that society does not have a right to protect its members by humane and effective measures.

- Acceptance of each other's personalities, cultures, tastes, preferences, differences and shortcomings.

- Appreciation of each other's talents, insights, and experiences.

- Respect for the rights of others as defined by the United Nations Declaration of Human Rights and the Canadian Charter of Rights and Freedoms.

- Respect and appreciation for our planet Earth and all her creatures.

- Dedication to live in a way that sustains our environment.

- Love for self and others including persons, things, life itself, and the All.

Nations, cultures, and societies more and more are being forced to learn to live together cooperatively. Like it or not we have become a global village facing an ever-dwindling supply of vital resources largely because of overpopulation and our abuse of the environment. Nor has the problem of the control of powerful weapons yet been solved. Until recently the inhuman side of nationalism was accepted as inevitable. Surprisingly, organized religion did nothing effective to curb such national violence. The hope is that now more of us know better.

Moral choices like other choices are possible in humans because of our intrinsic imaginative ability. I believe that imagination is behind everything that makes us human. And like other abilities it can be used well or badly. As anthropologists

Chapter 2

and historians have discovered, it is hard to find any human culture that does not acknowledge some form of the Divine in its collective imagination. The assumption that there is a positive force in the Universe greater than humans seems to be a requirement for human happiness and ethical development. Some would call this faith. I see it as creative imagination.

Although there are problems with too much faith, too much doubt is also a problem. Many people today seem to suffer from a habitual attitude of doubt. They doubt themselves, their loved ones, their government, and society in general. Let's hope this is temporary and due to the fact that we're all in a period of transition from many old, rigid, sectarian views of life to a new, dynamic, universal view of life based on our common humanity and scientific knowledge.

In a period of rapid and accelerating change it is also harder to find a moral compass. People tend to give up trying, and refuse to commit themselves to any ideal. It is quite understandable why many people are confused, fearful, cynical, and demoralized. But we have also become more aware of ourselves due to new insights that psychology and sociology now offer us. This new consciousness can help us make better choices. With patience and persistence we need to work at being more honest and forgiving of others and ourselves. The ability to laugh at oneself is important too, and not to take little things too seriously. A sense of humour also helps in the daily give and take of human relationships. As highly individual as humans have become, yet we are the most social of creatures and must pay close attention to social relationship and ethical process, especially when apparent conflicts arise. Not being sure of things is no excuse for not making the best of things.

Today we desperately need to learn not only how to respect cultures other than our own, but to cooperate with them in solving common problems. Do the Abrahamic religions, for instance, offer useful examples of peaceful coexistence? Robert Wright in his book, *The Evolution of God*,[13] points out that though there were some periods in history when Judaism, Christianity, and Islam tolerated each other, conflict between them has been more common. It seems to me that unless such traditional religions do the hard thing and give up their claim to absolute and complete truth, they are unreliable guides for an ethical compass acceptable to the world community.

So, if nature, religion, and science are not completely reliable guides as to values and ethics, what to do? Dare I suggest the following?

- Try to discern the merits of every case.

- Pick and choose values and ethics from nature, religion, and science as seems appropriate for each case. Use a consensus process. Include the wisdom of ethicists and the insights of artists in this process.

- Look for guidance from one's own conscience according to these values.

- Discuss ethical questions with a respected individual or group that has been devoted to these matters. For me, this would be my Quaker Meeting.

And yet, it seems to me there is something more. It is an awareness of the Other, the larger scope of the Universe of which humanity is only a tiny part. This is something so easy to forget, preoccupied as we are with our individual and group activities. But to ignore the Other is to be lost. The ancients called this forgetfulness "hubris" or pride. The tragic consequences of hubris were told by the classic Greek authors in many a tale of kings, queens, and heroes, but are really about us "writ large." To me, hubris includes our subconscious self-centredness. The price we pay is banishment, not from Eden but from the real garden of life on this planet with its joys and sorrows, its life and death. Extremists on both sides of the science-religion debate take note! Believing that only your group possesses the whole truth is hubris!

Summing Up

I believe most of us most of the time seek a deeper, more satisfying relationship, not only with each other but also with the rest of creation. I also believe in a paradox: God continues to create us through time and we humans continue to create God in our imagination. Religion is nothing more or less than a communal artwork. Religions as collective works of art are similar to languages; they offer interpretations of reality based on the experience and values of a particular time, place, and people.

As long as we don't confuse what we imagine with what is real, we will be fine. Believing in something at a basic level is a psychological necessity in everyday life. It is as natural as breathing. How can I sleep tonight if I don't believe that dawn will break tomorrow morning? Above all it should be love, not fear, that inspires belief.

In their search for objective truth, scientists have learned that it is best to keep the human element out of their experiments and calculations. Compared to the unimaginably large and ancient

Universe, human beings appear very insignificant (by "Universe" I mean God as matter/energy). Yet it seems that questioning beings such as us are being asked by the Universe to approve and assist Its evolution. Could it be that the Universe somehow needs and desires our cooperation?

I imagine that we as human beings are called to invest our energy and love in the great enterprise of life and existence. This we must do in a deliberate manner before we can begin to "have a life." It is as if God were asking, "Help me evolve." How could we refuse such a call? How could we ever think life was not worth living or that human beings were not a valuable part of the whole?

It is clear to me that we humans need to form some idea of the Other in terms that relate to us personally. Living as all of us do in a global village, "Others" now more than ever include our neighbours next door as well as neighbours a continent away. But science is wary of this task, rightly preferring its invaluable role of presenting the truth to us as impersonally as possible, leaving it up to philosophers and theologians to find (or invent) the human meaning behind the facts. (The better artists actually do most of this work when they can.)

Of late it has become difficult for either philosophy or religion to function with wide success, perhaps due to our unsettled times. What to do? As I suggested earlier, the groundwork in developing a worldview that really reflects our present situation might well fall on the shoulders of non-professionals. We could do this by delving into our hearts and souls. Then, by using our individual and collective imaginations, we could develop concepts and stories that we could then share and compare. Responding to these initiatives by us at the grass roots, professionals might later contribute their experience and expertise to this effort.

Chapter 2

As for God, though I have referred to Him, She, or It as the Other; God is also the All, which includes us. The All is eternal but is in constant motion, like a vibration between its expansive, creative phase and its contracting, destructive phase. This goes on at countless different levels from the subatomic to the cosmic. In living beings the motion of the All is manifested by their birth and death, and by the life and death of the billions of cells in their bodies throughout their lifetimes.

There is only one world, the physical universe. It has two apparently contrary characters - that of nature and that of imagination. It is very difficult for humans to imagine God in the abstract. So, long ago narrative timeframes were constructed such as the Christian one that consists of creation, fall, redemption, and second coming. Today we have come to realize that we project our personalities on gods or God and create these stories to make sense of things. Nothing wrong with that provided we remember that it is we, not God, who does all this! The why of Creator and Creation remains essentially a mystery.

Weird or What?

Accordingly, I would like to close this essay by sharing one of my theological/cosmological fancies. It may be similar to the notion or idea of others. On the other hand, it might appear quite weird and eccentric. But for me it is more believable than the miracles and creation stories of most religions, and more fun than most modern philosophy.

Here is my version of the divine narrative: Before the Big Bang, God was pure spirit and dwelled only in an unreal world, the World of the Imagination where He had been for eons. (I think of this God as male, because no woman would have been as impractical as He.) He amused Himself by imagining all the real worlds that could exist and how they might evolve in different ways. At first He was quite content being perfect, changeless, almighty, all knowing, and all the other impossible things God is supposed to be. In short He was unreal and boring – but he did have potential.

Eventually, God became lonely. His loneliness got the better of Him, and He acquired some new virtues He hadn't needed before: courage, hope and love. It also made Him decide to become real. To do this, He had to take His first leap of faith by sacrificing all His godly qualities except his eternal existence. He had to love and marry a beautiful but independent-minded lady called Matter/Energy. So He joined her in a terrifying ceremony called the Big Bang in the Church of Time! Because their marriage was so complete God became a hermaphrodite, acquiring a female character while remaining masculine. The rest is history.

God's greatest fear, though, was that She/He would lose the World of Imagination that He/She still needed as much as ever. But as we all know but sometimes forget, Imagination did not disappear, and, at least on planet Earth, creatures of imagination have been evolving since life began. After humans first appeared, fantastic stories of the imagination have not been lacking. God and Matter/Energy thus became parents countless times over. Not always very good parents, it must be admitted, but some of their kids survive and do quite well in the real world. And what a good thing the World of the Imagination was never lost!

Chapter 2

Addendum

I have found the following quotes especially helpful in seeking to understand the complementary roles of science and religion in today's world.

> *Different beliefs can open the mind to possibilities previously undreamed of, and this open-mindedness can be best achieved by maintaining a compassionate dialogue between all sides of the spiritual debate, especially between scientific and religious views. I believe this is what Einstein was suggesting when he said "science without religion is lame, and religion without science is blind." Whether we are gazing through a telescope, or contemplating our soul, we can all marvel at the beauty and mysteriousness of the universe.*
>
> Andrew Newberg, *Why We Believe What We Believe*[14]

> *Our minds have been built by selfish genes, but they have been built to be social, trustworthy, and cooperative. That is the paradox that this book has tried to explain. Human beings come into this world equipped with predispositions to learn how to cooperate, to discriminate the trustworthy from the treacherous, to commit ourselves to be trustworthy, to earn good reputations, to exchange goods and information, and to divide labour.*
>
> Matt Ridley, *The Origins of Virtue*[15]

Nonzero-sumness is a kind of potential. Like what physicists call "potential energy," it can be tapped or not tapped, depending on how people behave. But there's a difference. When you tap potential energy – when you, say, nudge a bowling ball off a cliff – you've reduced the amount of energy in the world. Nonzero-sumness, in contrast, is self-generating. To realize nonzero-sumness – to turn the potential into positive sums – often creates even more potential, more zero-sumness. That is the reason that the world once boasted only a handful of bacteria and today features IBM, Coca-Cola, and the United Nations.

Robert Wright,
Nonzero: The Logic of Human Destiny[16]

At the core of every religion lies an undeniable claim about the human condition: it is possible to have one's experience of the world radically transformed... The problem with religion is that it blends the truth with the venom of unreason...But a more profound response to existence is possible for us, and the testimony of Jesus and of countless other men and women over the ages attests to this. The challenge for us is to start talking about this possibility in rational terms.

Sam Harris, *The End of Faith*[17]

Chapter 2

> *There seems to be a steadily shifting standard of what is morally acceptable. Donald Rumsfeld, who sounds so callous and odious today, would have sounded like a bleeding heart liberal if he had said the same things during the Second World War. Something has shifted in the intervening decades. It has shifted in all of us, and the shift has no connection with religion. If anything it happens in spite of religion not because of it. The shift is in a recognizably consistent direction, which most of us would judge as improvement.*
>
> Richard Dawkins, *The God Delusion*[18]

> *When I was a Marxist ... I did have the conviction that a sort of unified field theory might have been discovered... [Marxism had] its messianic element...that an ultimate moment might arrive... But there came a time when I could not protect myself... from the onslaught of reality... the very concept of a total solution had led to the most appalling human sacrifices, and to the invention of excuses for them. Those of us who had sought a rational alternative to religion had reached a terminus that was comparably dogmatic. There are days when I miss my old convictions as if they were an amputated limb. But in general I feel better, and no less radical, and you will feel better too, I guarantee, once you leave hold of the doctrinaire and allow your chainless mind to do its own thinking.*
>
> Christopher Hitchens, *God Is Not Great*[19]

Faith Behind Faith
Tracking Down Ecological Guidance
Keith Helmuth

Starting Down the Trail

Among the persistent interests of my childhood was the search for arrowheads. My brother and I regularly walked the freshly ploughed fields near our home in northeast Ohio watching for that tell-tale shape offering clear evidence of the deep human past. Harlan, five years my senior, was well versed in woodland skills and historical lore. We had a keen awareness of the original people who had lived here before us between the Cuyahoga and Chagrin Rivers, inland from the south shore of Lake Erie.

Harlan was adept at spotting arrowheads; I was not. I seemed to lack the visual concentration needed for success in this quest. Watching the clouds was more my natural inclination. Then one day walking home from school, I took a shortcut through a neighbour's cornfield; there on the open ground between the edge of the woods and the first row of corn was an arrowhead. I was overjoyed! My first!

Chapter 3

I can still remember the feeling of picking it up. I wiped it off, polished it against my pants, and then realized it was peculiar. Instead of being the lustrous gray flint of typical arrowheads, it was a variegated combination of milky-rose and reddish-brown. It was about half the size of regular arrowheads and of a rough, irregular cut. But no mistake – it was definitely an arrowhead.

That evening after milking and while we finished up our barn chores, I told Harlan I had something to show him. I put the arrowhead in his hand. He looked at it very carefully, turned it over and rubbed his finger along the cutting edge. He studied it some more and then said something that lodged in my mind and has never let me go. "It was probably made by a young Indian boy who was still practicing at making arrowheads."

The image of that boy – probably about my age – rose up before me. I thought about him when I woke up the next morning. I thought about him in school. I thought about him on Sunday in church and I knew from that time on – knew of a certainty – there was something incomplete about my religious heritage. That arrowhead was incontrovertible evidence of a truth not accounted for in any Bible stories I had yet heard.

That aboriginal boy flaking out his rough arrowhead, and the whole cultural world to which he belonged, was a truth that stood outside the whole of Christendom. From that experience, my woods-wandering feet were set on a particular path of inquiry. I did not reject my Christian-Anabaptist-Mennonite heritage, but I knew in my bones there was a larger tradition, a broader story of human culture on the earth that had to be taken into account, and with which I felt a deep affinity.

Gaining Focus

Years later, at the State University of Iowa, I became interested in the emerging science of ethology - the study of animal behaviour - and the adaptation of species to various and particular environments. It seemed to me it would be useful to look at human cultures and environmental adaptation in a similar way. One evening, during a conversation with a professor at the bookstore where I was employed, I mentioned my interest in the study of animal behaviour with a view to shedding light on human behaviour. I got a sharp negative reaction. I was told that human beings are creatures of reason and culture and animals are creatures of instinct and conditioning, and there is no common measure between them. Period!

This professor was a learned and, supposedly, wise man, but I knew what he had said was incorrect. I had grown up in close association with various kinds of animals, both domestic and wild, and I knew that learning and behaviours that could be called cultural were clearly evident in many species. I was already well enough acquainted with the study of animal behaviour to know there was a professionally credible basis for my interest. I was working my way toward areas of study we now call historical ecology, social ecology, and sociobiology.

By this time it was clear to me that the story of my own culture was embedded in the wider story of human culture in general, and that the human story was embedded in the larger story of the primate and mammalian world, which, in turn, along with all other forms of life, was embedded in the whole land community – the biosphere. And I understood that the biosphere was part of the still more comprehensive story of earth's evolutionary unfolding. I let the remarks of my professor pass uncontested and continued with my studies.

Chapter 3

A few years later, in the library of the College of Forestry at Syracuse University, I read Clark Wissler's pioneering monograph, *The Relation of Nature to Man in Aboriginal America*.[1] This study put a new light on my path. It helped focus the lens of adaptation through which I was more and more seeing and understanding the human-Earth relationship. Shortly thereafter, I discovered the work of the American geographer, Carl Ortwin Sauer. I became immersed in his collection of classic studies, *Land and Life*.[2] Encountering this master of historical geography and human ecology turned the light on my path into a sunrise, illuminating the whole of human history in relation to the environments of Earth in just the way I had been sensing was needed for a fully rounded ecological understanding.

This essay details some of the findings that have come to me as I have worked to develop an ecological worldview and track down the guidance it provides in relation to the human crisis of our time – a crisis manifesting in both the human spirit and the human-Earth relationship.

The Ecology of Faith

While the study of adaptation usually concentrates on resources, population, technology, and economic behaviour, it also encompasses culture in general and religion in particular. There is an ecology of faith at the core of cultural adaptation.

Religious faiths are fundamentally strategies of adaptation. They evolve and devolve. Some work better than others. Some gain in efficacy and others fade at different points in their trajectories. But they all exist and function in an ecological context. This context can be thought of as a kind of faith behind faith, a kind of primal grounding that underwrites the human enterprise.

In religious discourse, faith is often taken to mean trust in the truth and efficacy of a particular story about how things have come to be the way they are, how things are going to work out, and what it all means. In these terms, there are a variety of particular faith stories. People often inherit one of these stories or, at some point, make a choice about which story makes sense to them. This is not, however, what comes to me when I reach for a fully rounded sense of faith. What interests me with regard to the ecology of faith is primarily the energy, the relationships, the sense of reality, and the growth of identity that makes it possible to have a faith of any kind.

It has long seemed to me that there is a kind of faith behind faith, a primal energizing, organizing, and motivating sense of being in the world that backgrounds and continuously underwrites all particular configurations of faith. It is an upwelling sense of positive response to life that makes it possible for us to creatively negotiate each day despite the vagaries of experience, and even the shifting persuasions of belief that may emerge on our path.

This dimension of faith is like a solar energy cell for the soul. It is like photosynthesis for the psyche. It is a gathering from the whole context of Creation that suffuses the entire expression of life, and is articulated in every living form. If we look with full attention, we can see this dimension of creative energy in all the forms and processes of Earth.

One of the most profound thinkers in the dialogue of religion and science has been the great French Catholic priest and palaeontologist, Pierre Teilhard de Chardin. In the later years of his life he became especially concerned about this primal dimension of faith. He said what concerned him most about the spiritual condition of modern societies, and alarmed him about the human future, was "a dying down of the zest for life."[3]

In the terms I am using, his concern was about the loss of the faith behind faith, the loss of that primal connection in which we are energized and animated by an *unquenchable urge to flourish*. When we consider this flourishing and look upstream, we can see it flowing from the mystery and totality of our incredible planetary home. When we look downstream we can see it flowing into the mystery of Earth's astounding biodiversity.

This surrounding mystery is the origin of our sense of the Divine. Something is going on, we are in the middle of it, this is our home place, but it is way beyond our ken. What we know most closely is an uplifting "zest for life," which, in its most fully rounded expression, includes the unfolding of gratitude and compassion. Meister Eckhart, the great Christian mystic of the late 13th and early 14th Centuries, had a succinct way of expressing this dimension of our human situation. He wrote: "The eye with which I see God is the same eye with which God sees me." And again; "If the only prayer you ever say in your entire life is 'thank you' it will be enough."[4]

The Ecological World View and the Faith of Ecology

A fundamental reorientation is taking place around the question of collective human security and wellbeing. Issues of justice, peace, ecological integrity, and even a continuing faith in the human enterprise are all converging on the reality of the human-Earth relationship and its expression in economic adaptation. The human future will be reasonably secure or disastrously disrupted, depending on the way economic life is arranged and carried on. The relationships that are generating resources wars, social triage, entrenched inequities, and ecological disruption are all focused in economic adaptation. Ameliorating these conditions and altering the human trajectory toward greater equity, security, and wellbeing requires rebuilding economic activity within the biotic integrity of Earth's ecosystems. This is the faith of ecology.

This is the broadest possible concern that can be placed before the conscience of all human communities. We might ask why this concern should be of particular interest for faith communities since it must be addressed at all social, economic, political, educational, and professional levels? The answer is simply that religious authenticity depends precisely on bringing the energy of love and the work of community to the broadest possible concern for the human future. This is ecology of faith.

To understand the ecological worldview and the faith of ecology we need to ask what are the main areas of information, knowledge, and experience that must be brought into focus? For the purpose of this discussion, I will identify and briefly describe four tracks. Each track is referenced to a person whose work has contributed in a significant and accessible way to the creation of the ecological worldview. Although the figures referenced are well known to many students of ecology and culture, it will be useful to present their contributions in a way that links their work into a composite understanding. Each of these figures has been adept at coining phrases that catch and communicate the organizing concept of their work. In the scientific track we have James Lovelock with "The Gaia Hypothesis." In the cultural history track we have Thomas Berry with "The New Story." In the ecology and economic adaptation track we have Barry Commoner with "The Closing Circle." In the human-earth relationship track we have Aldo Leopold with "The Land Ethic."

Chapter 3

The Gaia Hypothesis. In considering James Lovelock's work, a clear distinction must be made between his formulation of the Gaia hypothesis and the subsequent adoption and promotion of the concept by others. The scientific work and scientific reasoning so ably recounted and illustrated in his book, *Gaia: A New Look at Life on Earth*,[5] has stood the test of almost three-and-a-half decades. It is this scientific work that is the focus of this discussion.

Through his experimental work on the interaction of chemical elements and compounds in Earth history and in the development of life, James Lovelock recognized a feedback and regulatory process. The history of this process helps provide an explanatory context for the persistence and flourishing of life within the environments of the planet. The evidence with which he was working led to a surprising and compelling conclusion: The evolution of the chemical composition of the atmosphere, and its increasing suitability for the flourishing of biotic process, could only be explained, in scientific terms, through the regulatory contribution of the whole biotic complex itself – the biosphere. The evidence indicated that once having gotten started, life, as a collective phenomenon, became a direct contributing agent to the maintenance of earth's atmosphere within a certain range of chemical composition – the very range, it turned out, required for the further development of life. And it is only through this continuing regulation of the atmosphere by planetary life, that planetary life continues to exist and is able to flourish with a high level of diversity.

The Gaia hypothesis provides ecological intuition with a comprehensive scientific context. People who were predisposed toward seeing Earth as a holistic process, responded with delight. Some elders from within Aboriginal cultural traditions responded with bemusement and a kind of patience tolerance. They said, in effect; "That's good medicine you have there. Too bad it took you so long to come up with it. Welcome to the Circle of Creation." Certain people who had always regarded Earth's environment as a stockpile of raw materials for human manipulation and consumption, became alarmed that their industrial ventures and the quest for endless economic growth and wealth accumulation could now be held to account against the history and science of biotic integrity.

Lovelock's scientific work provides a comprehensive context for the study of ecological relationships. It sets all life communities, including the human, squarely within the history of Earth process, and shows them to be entirely beholden for survival to the continuing integrity of Gaia – life process at the planetary level.

The New Story. Thomas Berry, a Catholic priest, trained in theology and the history of culture, came to regard himself as a "geologian." After a long life in the scholarship of religion and culture, Berry developed an understanding of the human story that brings the human-Earth relationship into focus. He sees the human-Earth relationship as central to the unfolding of culture, and all the facets of guidance, adaptation and behaviour that culture encompasses.

Berry observes that modern Western cultures are in a state of confusion with regard to guidance and adaptation, and are destructively floundering with regard to the human-Earth relationship. The story of human origin, cultural development, and moral orientation that has been built up out of the Judaic-Christian and Greco-Roman contexts has become seriously dysfunctional. Individuals and subculture groups may still organize their lives and behaviour according to some version of this "old story," but in its larger public and cultural dimensions it is failing to provide adequate guidance.

Among the most notable examples of this failure is the contemporary state of the human-Earth relationship. Berry notes this cultural failure as an autistic-like blindsiding of the organic circumstances of our lives and of Earth's biotic processes in general. The Western narrative has not engaged the human-Earth relationship in a way that offers adequate guidance. Instead, it has spawned a dominion story that now provides the only comprehensive guidance taken seriously at a public level in modern societies. This is the narrative of technological domination, maximum resource exploitation, unfettered capital accumulation, and unlimited economic growth. This story is promoted, and largely accepted, as the only reasonable scenario for the human-Earth relationship.

Thomas Berry describes an alternative. He sees a "new story" and a new sense of guidance in a composite narrative of scientific cosmology, evolutionary biology, global cultural history, and ecological knowledge. He combines the story of Earth as it has emerged from cosmic process, the story of life as it has emerged from Earth process, and the story of the human as it has emerged within Earth's biosphere. He sees ecological understanding emerging from both scientific work and an increased awareness of the beauty and diversity of Creation. He introduces this "new story" in his book, *The Dream of Earth*.[6] In *The Universe Story*[7] – written with mathematical cosmologist, Brian Swimme – he presents the whole sweep of cosmic unfolding, Earth history, and human emergence. In *The Great Work: Our Way Into the Future*,[8] he details how the new guidance we need emerges from ecological understanding and leads to a "mutually enhancing human-Earth relationship."

Faith Behind Faith

The concept of a "mutually enhancing human-Earth relationship" is one of Tom Berry's ecological guidance masterstrokes. It is a transforming conceptual advance over the conventional dualism that sees humans on one side and "nature" on the other. As Berry reminds us, the context is always the biosphere and biospheric relationships. The human-Earth relationship is our primary reality. "Mutually enhancing human-Earth relationship" expresses precisely the dynamic to which ecologically sound adaptation aspires. Berry is also responsible for introducing the concept "Earth process" into contemporary discourse. This is a functionally helpful replacement for the concept of "nature," which is made up of a confusing array of quasi-theological cultural constructions. The process of the planet, in all its diverse complexity, is what we are dealing with, not some metaphysical unity called "nature." Berry's clarifying concepts have been of cardinal importance to the development of the ecological worldview.

While Lovelock speaks mainly to the scientific track, Berry incorporates the scientific into the story of culture and represents the human as a constituent part of Earth's revelatory emergence and unfolding. Berry's work honours the scientific-cultural dimension and the religious-cultural dimension in the same discourse, and has become a principle guide for the cultural track feeding into the ecological worldview.

The Closing Circle. In the mid 1960s Brian Hocking, a well known Canadian biologist of the time, published a book with an arresting title – *Biology or Oblivion: Lessons from the Ultimate Science*.[9] He argued that the trajectory of our society's industrial-commercial adaptation is in serious conflict with the way the organic world actually works, and if we persist in this conflict, we

Chapter 3

are bound to crash our civilization. The book was issued by a small publisher, received little attention, and rapidly disappeared from view.

Less than a decade later, Barry Commoner, also a biologist, published a book that picked up on Hocking's theme. *The Closing Circle: Nature, Man, and Technology*[10] was issued by a mainstream publisher, received major attention, and became a prime text of the emerging environmental movement. As a professional researcher and educator on the physiochemical basis of biological process, Commoner is especially qualified to address the fundamental conflict between biospheric integrity and the technology of our economic system. He points out that behind the form and functioning of Earth's biotic environment there is, so to speak, two to three billion years of "research and development."

As a way of understanding the intervention of modern technology into this context, he offers a striking analogy. If you open the back of a fine Swiss watch and poke a sharp pencil into its works, there is an infinitesimal chance you will improve the functioning of the timepiece. The probability is much greater, of course, that the watch will be damaged. The watch is the result of a long tradition of highly skilled craftwork, and is not likely to be improved by such intervention. From the standpoint of biological systems, the modern capital-driven economy is wielding its technology in a similar way, with predictably disruptive and damaging consequences.

Barry Commoner was among the first to apply biological systems analysis to the dilemma modern economics has created within the human-Earth relationship. This dilemma is clearly illustrated by the fact that in order to maintain the capital-driven economy under present conditions, it is necessary to increasingly damage the functional integrity of Earth's ecosystems, and the biosphere as a whole. From the standpoint of science, this situation is devolutionary; from the standpoint of enlightened humanism, it is absurd; from the standpoint of religion, it is blasphemous.

Commoner's analysis of this dilemma is based on the "four laws of ecology":

- Everything is connected to everything else.

- Everything must go somewhere.

- Nature [Earth process] knows best.

- There is no such thing as a free lunch.

At first glance, these statements may appear simplistic, but they are solidly rooted in biological knowledge and in the thermodynamics of energy and matter. Taken together, they describe the ecological worldview and offer guidance for an ecologically based economic system.

In a second book, *The Poverty of Power: Energy and the Economic Crisis*,[11] Commoner develops a schematic formula that is both profound and memorable. Human settlements and social order depend on the operation of three great, interrelated systems:

- the planetary ecosystem,

- the human production system,

- the monetary exchange system.

Chapter 3

Ecologically speaking, the interrelationship of these systems goes like this:

- The planetary ecosystem is the source of all materials and energy processes that support human life.

- The production system is the network of agricultural, industrial, and service activities that convert earth's materials, energy processes, and relationships into the wealth that sustains human settlements and social life.

- The monetary system represents the value of this wealth in ways that facilitate its exchange. It governs how this wealth is distributed and what is done with it.

In an ecologically sound arrangement of these three systems, the governing influence would flow from the ecosystem, to the production system, and then to the monetary system. The continuing integrity of the ecosystem would determine the design and operation of the production system. The stability and good service of the production system would determine the design and functioning of the monetary system.

Our contemporary economic reality, however, has the relationship of these three primary systems exactly the wrong way around. The monetary system drives the production system into unlimited, consumption-based economic growth. The production system, in order to meet this demand of the monetary system, generally operates without regard for the health and integrity of the ecosystem.

The governing influence is flowing the wrong way, and the environmental crisis is the result. These comparative relationships can be diagrammed as follows:

Governing Influence	->	->	Outcome
Monetary System ->	Production System ->	Ecosystem ->	Ecological Breakdown
Ecosystem ->	Production System ->	Monetary System ->	Ecological Health

Barry Commoner's formula clearly illustrates the science-based approach to an ecological sound economy, and the policy issues that must be addressed on the way to an ecologically sound way of life.

The Land Ethic. And lastly, consideration must turn to the founding figure of modern ecological consciousness – Aldo Leopold. Leopold was a conservation biologist whose work encompassed field research, university teaching, public policy, and philosophical reflection that holds up the ecologically embedded basis of ethical development. He had the ability to frame his thoughts and insights in plain, memorable language. His best-known book, *A Sand County Almanac*,[12] published in 1949, collects his sketches from the field and his reflections on humanity's relationship with the land community. One would never suppose from such a modest title that this book would become one of the prime sources of ecological consciousness in our time. Leopold's skill was twofold: He articulated a philosophy of ecology in a language of such quiet beauty that we get not only the conceptual understanding, but also the experience of the spirit in which he lived and worked.

Chapter 3

In *A Sand County Almanac* he argued that the recognition of the "land community" is the preeminent discovery of modern science. This may seem a curious claim when such an array of dramatic discoveries, especially since his time, could be named to this honour. But if we think carefully about this, I believe we will see he is correct, and will continue to be correct for as long into the future as we care to imagine. The scientific recognition of the "land community," and its ecological integrity, is the fundamental context of human adaptation and wellbeing. The same cannot be said for any other context of scientific discovery.

Leopold suggested the next major step in the evolution of human moral sensibility would be the development of "the land ethic." He offered this formulation: "A thing is right when it tends to preserve the integrity, stability and beauty of the biotic community. It is wrong when it tends otherwise." Many volumes have since been written on the philosophy of ecology, but it is this simple statement, with its emphasis on the aesthetic factor in moral awakening, that has become the touchstone of the ecological worldview.

In Summary. James Lovelock describes the emergence of life as an expression of Earth-process, an expression that is characterized by a homeostatic regulatory function within biotic development itself that maintains the chemistry of Earth's environment in the very condition required to enable the flourishing of life to continue. Understanding this ecological relationship inducts us into a great responsibility – the responsibility of being co-workers in the maintenance of the commonwealth of life.

Thomas Berry describes the cultural context of this relationship, and details the range of activities that flow from the exercise of this responsibility. He calls these activities "the Great Work."

Barry Commoner describes the process and relationships that compose the organic world. He explains why the capital-driven market economy is deconstructing ecosystem integrity and cannot be sustained. He describes the ecological orientation toward economic adaptation.

Aldo Leopold describes the enhancement of the human-Earth relationship based on the emergence of "the land ethic." The land ethic, according to Leopold, comes into full effect when scientific knowledge and aesthetic experience of Earth and its life communities rise into reverence, respect, and love.

This is the point at which science, culture, economics, and the human-Earth relationship converge into the ecological worldview, and the ecological worldview becomes the expression of authentic, revelatory experience – the ecology of faith and the faith of ecology. This experience, in its most fully rounded expression, unfolds with a sense of presence that calls us to wake up within the greater life of beauty, service, and love. If we can collectively, and globally, take up this ethic and become responsible citizens of the Earth community, a mutually enhancing human-Earth relationship may yet be achieved.

Faith Behind Faith: Steps to an Ecology of Practice

Within the panorama of human cultures and behind the particularities of each culture's story of faith, there is another story, another level of deep faith, a background context of energy and relationship that animates human experience and nourishes creativity. I have a sense of this faith behind faith as a fluorescence of the spirit, as an incandescence of the soul. It is the energy and creative orientation of this deep background faith that is the

Chapter 3

funding source of culture, and that enables us, within our cultures, to create our particular stories of faith.

This faith behind faith is a gift, not a mental construction or theological exercise; it is a gift given not only at the human level, but into Earth-process as a whole, and most notably, into Earth's biotic process, manifesting in every form of life as an unquenchable urge to flourish. It is as simple as that. It is also the great mystery. On the human side, it underwrites the scripts of culture in which various images, metaphors, story lines, and systems of symbolic representation and meaning become expressions and amplifiers of faith.

The faith behind faith is not looking for allegiance; it is looking for expression. It is not the song, but the signal - the pulsing energy that animates. It is both before and after words. It is the most illusive, yet the most intimate breath of things that draws us up into the warp of life and out into the weave of the world. It gives us a way of working for the good of all, even if we must go through the worst of times.

Whatever collective, public policy choices on the human future are made, and then worked out over the next several decades, it seems likely that most of us currently alive, and certainly our descendants, are in for a journey of re-adaptation that will test the resilience of faith. As to what specific, practical steps can be taken to build up the resilience of the faith behind faith, here are four suggestions.

The Metabolic Step. The metabolic factor in the maintenance of faith is rarely considered, but it can be critical. The cells of brain tissue are nourished in exactly the same way as muscle tissue. Everyone understands that muscle tissue cannot function normally if it is inadequately nourished. The brain, likewise, cannot function efficiently if the cells of its tissue are lacking the full range of critically essential nutrients. The brain is the seat of consciousness in general and thought processes in particular, factors that are central to the operation of faith in our lives. Yet, almost no attention is given to brain nutrition by the scholars of faith. The same observation can be made with regard to the endocrine system – the seat of emotional response and balance. Emotional response, emotional balance – or lack thereof – are important factors in the functioning of faith.

The neurological and endocrine systems, in particular, are the context in which the experience of faith emerges, and through which we develop and extend our spiritual life. This is not a startling insight. Our metabolic situation is the only house we have. Nutritional shortfall and metabolic inefficiency directly affects our ability to function in all ways, including the spiritual. The emergence of a sustaining, primal faith – the faith behind faith – is, in part, a matter of nutritional intake and metabolic efficiency. The fact that neurological and endocrine processes are subject to considerable variation due to genetic and environmental factors makes attention to nutrition and metabolic efficiency a frontline step in the ecology of faith. Appropriate nutrition underwrites the zest for life and supports the faith behind faith.*

---

*Friedrich Nietzsche, the renowned German philosopher, became aware of the importance of the "metabolic step" near the end of his life and wrote the following lament: "I am interested in quite a different way in a question upon which the 'salvation of mankind' depends far more than it does upon any kind of quaint curiosity of the theologians: the question of nutriment. One can for convenience sake formulate it thus: 'how to nourish yourself so as to attain your maximum strength?' ... My experiences here are as bad as they could possibly be; I am astonished that I heard this question so late ...Until my very maturest years I did in fact eat badly. ... With the aid of Leipzig cookery, for example, which accompanied my earliest study of Schopenhauer (1865), I very earnestly denied my 'will to live.' ... German cookery in general – what does it not have on its conscience!" *Ecce Homo*, translated by R. J. Hollingdale, 1979. New York: Penguin Books.

Chapter 3

The Metaphysical Step. The thorniest metaphysical problem in the whole of human experience – the problem that gnaws deep into the marrow of faith and raises the temptation of fatalism – is the question of evil. Much theological ink has been poured over this problem, but it is not a theological solution that is required. On the contrary, as George Fox, the founder of the Religious Society of Friends (Quakers), discovered, an experiential solution is needed to release the paralyzing hold of this problem. In Fox's case it was experiencing a vision of the "ocean of light" overflowing the "ocean of darkness" that accomplished his release.[13]

Within the structure of human perception and mental functioning there is an unremitting dualism that continually foils our great desire for unity. We know without doubt the goodness of many things, but, at the same time, evil has a pattern of recurrence that keeps us on the metaphysical rack. The structure of human knowledge and the processes of moral reasoning are constituted in such a way that a comparative element is always at work in the way we come to know and understand things. But we can take a further step and observe that this comparative dynamic has a characteristic substructure: In the dance of opposites, the positive always sets the stage and leads the performance. As bad as things may seem to be, evil can only emerge and take shape against a background of goodness.

So even in our darkest moments, even when we are confronted by what seems to be the absence of goodness we have this insight to rely on: Absence can only be known in relation to the experience of presence. A sense of absence cannot even begin to emerge without the reality of presence. Likewise, evil can only be known within the larger reality of goodness.

This realization does not magically remove the experience of anguish or exempt us from any moral task with respect to evil, and it may not be helpful at all for some folks, but it is the experiential moment of understanding to which many struggling souls have come. If not a complete resolution, this perspective is at least a resting place for the metaphysical problem of evil - a resting place that allows the faith behind faith to catch a better stride for the journey.

The Social Step. As important as it is to get the metabolic step on track and the metaphysical step into perspective, they are mainly background to the full emergence of primal faith in the social context of our lives. Human associations are the richest and strongest support for the emergence and maintenance of faith. The special place of communities, whether religious or not, is of particular importance. Participating in community-based associations is the terra firma of the faith behind faith. This is the step and the context that is most self evident and familiar to us.

There is, however, another dimension of the social that is also important but which is often overlooked. The entire biospheric realm is, in every detail, an intensely social phenomenon. Energies, relationships, reactions, and responses are flowing back and forth, up and down, and crosswise everywhere throughout the weave of the world. From the microbial composition of soils to the great flocks of migratory birds, from the ocean-cruising pods of whales to forest tree succession, from the companionship enjoyed between animals and humans to the honey bee scout communicating through dance-like movements to the rest of the hive's workers where the best nectar harvest of the day can be found, a vast tapestry of social relationships compose the biosphere: All this, and everything in between, reveals relationship as the primary context and expression of life.

There is no end and no "outside" to the sociality of Earth. Human social order is embedded in and completely dependent

Chapter 3

on the larger domain of relationships that make up Creation as a whole. The extent to which we are knowledgeable about these relationships, and take human embeddedness in this larger social realm into account, the greater will be our adaptational integrity and resilient functioning within Earth's commonwealth of life. The more the scope of these relationships is realized in the practical details of our lives, the more our communities will become places of ecological and social coherence, and thus foster and nourish a fully rounded and flourishing faith behind faith.

The Ecological Identity Step. In her important book, *The Ecology of Imagination in Childhood*,[14] Edith Cobb focuses on the significance of experience in the natural world for cognitive and creative development. Her research reveals a common pattern in the development of emotion and imagination in children. It goes like this: At some point before the age of 10 or 12 an experience, or perhaps a series of experiences, with some aspect of the natural world gives rise to a sense of beauty and mystery, wonder and awe.

Such experiences may occur in the close presence of animals, or when under certain trees in a special place in the woods, or when watching the moon rise and turn a dark lake to silver. Such experiences can come just watching a hummingbird at a feeder, or spotting a red-tail hawk in a city park, or watching the endless rolling of the ocean. Gazing into the night sky bright with stars, or out over landscapes - either well known and comforting, or new and mysteriously beautiful - can provide deeply formative experiences. All these experiences, and many others of a similar sort, call up a sense of communion and open a window in the soul. When the heart and mind go out and enter a part of the larger natural world, and when that part of the larger world, in turn, enters one's life and becomes, in effect, a part of one's identity, a sense of deep natural connection, of solidarity, settles in the soul.

This kind of experience can become a life-long source of intuitive understanding. It can provide a deep and ongoing sense of affinity. It can serve as an aesthetic reference point, an ethical compass, and a guide to compassionate moral action. Such experiences open a path of development into ecological consciousness and ecological identity; they comprise a relationship with the primal; they give rise to the faith behind faith.

Although these experiences seem to occur most readily in childhood, they are by no means confined to that period: They can also be cultivated in adulthood. The extent to which adults are now seeking out these experiences may be judged by the fact that researchers in developmental psychology are now studying what they have named "ecological conversion." The ecological identity step transports the imagination into the consciousness of communion and helps strengthen the faith behind faith.

These four areas of attention do not complete the steps to an ecology faith, but they are something of the matrix from which the faith behind faith develops and is sustained. In this context, it may also be asked what has happened to the theological structure of belief that is conventionally associated with religious faith? The answer is, nothing in particular has happened to them. Depending on our cultural circumstances, they may maintain the same significance, or they may evolve into new configurations. The point of this exploration is not to replace the particularities of one faith with another, but to recognize and suggest ways of nourishing the faith behind faith. The steps outlined here constitute a practice that can be usefully allied with various cultures of faith. They can help avoid the dead end temptation of fatalism and, hopefully, better equip us with ecological guidance to weather the great difficulties that are likely to come.

Chapter 3

What Does Creation Have in Mind?

We are in great need of ecological guidance. Teachers of indigenous wisdom tell us to remember our "original instructions," our intuitive understanding of our place in Earth's commonwealth of life, and how to live with integrity in our home place. Ecological science now provides substantial confirmation that indigenous wisdom is on the right track. What *does* Creation have in mind? Faith behind faith comes back again and again to this question, the question of how to live in a mutually beneficial relationship with the integrity of Creation, the question of "right relationship."[15]

It is not unreasonable to wake up in the morning and wonder how long our modern civilization can go on like this. How long can the capital driven, industrial-consumer economy continue its unremitting growth? Can it go on forever? If not, what happens when it stops? Or what happens, if before it stops, the quest for maximizing wealth destroys the natural health and balanced functioning of Earth's ecosystems on which human communities ultimately depend? If we take the integrity of Creation - the integrity of Earth's ecosystems - as the fundamental context of the commonwealth of life, it is not hard to see that our current industrial-consumer economy is steadily degrading the planet's ability to support life, and, if carried to its logical conclusion, will end in ecological, economic, and societal collapse. Exponential growth on a finite planet simply won't work.[16]

From the standpoint of science, this is a devolutionary situation – the unraveling of Earth's biotic complexity. From the standpoint of an enlightened humanism, it is absurd. From the standpoint of religion it is blasphemous. From the standpoint of the faith behind faith it is the "dying down of the zest for life."

Faith Behind Faith

How have we gotten into this situation and how can we get out of it? How can we bring the ecological guidance that both science and the faith behind faith provide into full effect?

Modern life is based on the assumption that the environment is part of the economy. This assumption is an error. The human economy is actually part of the environment – a wholly owned subsidiary of Earth's larger biotic and geochemical functioning. This recognition is a profound upheaval in our culture's understanding of the human-Earth relationship. For those who think in theistic terms, it means a significant theological reassessment of humanity's place in the larger context of Earth process. For those who think in scientific terms, but are still wedded to the notion of human dominion, the same reassessment of the human-Earth relationship is required. The Earth sciences in general and ecological science in particular, require a worldview in which the human-Earth relationship is understood to be a continually emerging and unfolding process of adaptation and reciprocity. This understanding of relationship goes to the core of human identity within Creation. It makes ecologically coherent adaptation a matter of spiritual significance and religious responsibility. It places economic and social life under the guidance of the integrity of Creation and brings the ethic of right relationship into clear focus.

When we ask the question, "what does Creation have in mind?" we are mostly asking what kind of economic adaptation is favoured with ecological resilience and sustainability. It may be difficult for mathematically trained economists to approach this question, but that just illustrates the difference between mathematics and ethics, and tells us why we need to bring the ethical perspective into the study and practice of economics – a perspective that will help economics be the kind of science it really is, a *social* science.

Chapter 3

The question, "what does Creation have in mind?" can be phrased in a variety of ways. Planetary scientists can ask, "What does the biosphere have in mind?" Ecologists can ask, "What does this ecosystem have in mind?" Educators can ask, "What does this school, this learning programme, have in mind?" City planners can ask, "What does this urban region have in mind?" Folks who live on flood plains *should* ask, "What does this river have in mind?" Foresters and woodlot owners can ask, "What does this woodland environment have in mind?" Farmers can ask, "What does this land, this climate, this market have in mind?" This question of right relationship can be applied to all vocations, employments, and habitations, and through the whole range of activities that constitute the adaptation of human communities to their various ecosystems.

Now if this seems like too much of a stretch, allow me to explain why I think this is really what is going on – or should be going on – as we try to figure out if the human project can be put on a more Earth friendly footing. Although the question can be framed in many ways and at different levels, it is always keyed to the human-Earth relationship. From the mystic and theologian trying to read the mind of God in Creation, to the city planner and home gardener trying to figure out what should go where, the question is basically the same: What kind of relationship and what kind of adaptation is appropriate for life's flourishing and the common good, for the continuing integrity of Creation? What is potential in this situation? What are the emergent properties, encompassing tendencies, and creative processes that are available to us as we move into, settle on, and work with any particular environment, either natural or cultural – or, as is usually the case, a combination of both. The context is always the human-Earth relationship, and the question is always how to make the relationship mutually enhancing.

Faith Behind Faith

I have been challenged on this line of reasoning by those who argue that nothing has anything in mind with regard to the way Earth works, that the entire cosmological context, including the human-Earth relationship, is wholly experimental. Some of these folks insist, as an article of "faith," that randomness rules, and that we are simply free to make of it whatever we will: To which, I sympathetically reply; But of course! Isn't that just the problem? Human economic behaviour is free to trash the environment, to destroy ecological and social integrity, but that in no way alters the context or question of resilient and sustainable adaptation. It only puts us on notice for wising up and staying alert to the realities of how the biotic processes of Earth's ecosystems actually work.

What does Creation have in mind? The full force of this question came to me over thirty years ago when I was in the middle of developing and managing a farm operation in as ecologically coherent a manner as possible. During this time, I sketched out the following meditation:

> *By great good fortune we live in place where, on clear nights, we can go out and gaze at the galaxy – the Milky Way – flowing like the River of God across the sky. From horizon to horizon this great wheel of light flows with the turning Earth and shifts its angle with the seasons. Tonight, its course has an east-west bearing and the other stars fall away to the north and south like the slow rolling wake of a great cosmic ship.*
>
> *Part of being a farmer in this particular place on Earth has included, for me, this midnight conversation with the Cosmos. Resting from the manure hauling of the day past, and thinking of the potato digging of the day to come, I take this lonely but comfortable time to once again ponder on what it all means: All this expenditure of energy, all this human cunning, all this*

‹ 95 ›

Chapter 3

seedtime and harvest, all this buying and selling. How does it all work? How do we keep it all going?

Down in the valley of the Saint John River I hear the steady drone of heavy trucks and speeding cars. For the moment, I hear it all from a distance, but so often I am in the middle of this energy-sucking caravan; restless souls and endless commodities flowing to and fro over the Earth. Around the hill to the south, large tracts of forestland are being clear-cut for newsprint, advertising, packaging material, and to keep the growth-driven, high-consumption economy rolling. Is this what Creation really has in mind for human communities and their relationship to the land?

When I gaze into the night sky and feel the great wheel of life turning, I am struck by the fact that most of what we experience as positive and progressive in our high-energy civilization, Earth's biosphere experiences as negative and retrogressive, as the breaking up and closing down of life support relationships. Surely this cannot be what Creation has in mind! To accept this ravaging of life as somehow normal or necessary is to disable the compass of faith. As surely as the River of God flows across the heavens and throughout the cosmic deep, we know it flows in nurturing and creative presence throughout life on Earth as well. We are of this Earth, the biosphere is our home, and if we better harmonize our economic activity with the way Earth's ecosystems actually work we can get ourselves on a better footing with the great flourishing of life that Creation has so clearly, so consistently, and so beautifully in mind.

Our "original instructions" come to us through the way of the heart, and the way of the heart leads unerringly to a sense of home. A sense of home is the origin of our sense of the sacred. Here we find our "original instructions" and the guidance we need begins to unfold. Only when we have a strong sense of solidarity with the whole commonwealth of life, and a strong sense of Earth as the home place of life's commonwealth, will we live and work for the common good with the love we naturally have for home. Our economic adaptation will then be devoted to creating a mutually enhancing human-Earth relationship, which, after all, is what tracking down ecological guidance is all about and what the faith behind faith is for.

Addendum

Quakers in Science

From *Quakers in Science and Industry* by Arthur Raistrick. Philosophical Library: New York, 1950.

The seventeenth century was a time of real advance in science and of the formulation and extension of scientific enquiry and experiment along modern lines. Alongside the religious questings and searchings out of which Quakerism emerged there was an ever-increasing urge to explore and to understand the physical world and its implications. There was a new acceptance of the function of observation and experiment in the search for an elucidation of the laws of nature which was a direct break away from the methods of the schoolman, and which called for a new type of scientific mind. The world had only recently been awakened to the potentialities of the new astronomy by the work of Galileo (1564-1642) and Kepler (1571-1630), and there was an eagerness to apply Galileo's new instruments, the telescope and microscope, to study the external world in new dimensions. Descartes (1596-1650) had introduced a philosophy which in every part of it was a challenge to established and customary thought. Contemporary with the first generation of Friends, Newton (1642-1727), who was to be closely associated with them through his official

Addendum

duties at the Mint, was laying the foundations of mathematical philosophy. It was impossible for people endowed with the active, enquiring spirit characteristic of Friends, keenly alive to the unity of life and dedicated to a way that was exceptional for the times in its searching out and love of truth, to stand apart uninfluenced by this stream of new knowledge. It is not, therefore, surprising that Friends should have had close contacts with, and even contributed to, this birth of modern science, in spite of the many serious obstacles that stood in their way. What is surprising, is that this aspect of Friends' activities should have passed with so little comment, although it has been stated in recent years that in strict proportion to their numbers, Friends have secured something like forty times their due proportion of Fellows of the Royal Society during its long history. Surely it is significant that the scientists and the Quakers share the insistence on the complete surrender to the guidance of truth, and that the scientist, if his science is to flourish, must exercise an ethical standard of integrity and unswerving loyalty to the revelation of truth that his work affords, that is identical with the faith of Friends, being expressed only in a different vocabulary.

In 1645 a group of persons [in London] interested in the new philosophy drew together and fell into the custom of a weekly meeting at which they could discuss the topics which were exercising their thoughts. The group had no rigid membership, nor were the attenders by any means constant, it was a friendly gathering to which people came as friends. Their material for discussion was ample, the works of Galileo, Kepler and Descartes, were almost fresh from the press, and the observations they made were being repeated with more or less success. Three years later a similar group was formed in Oxford and a close contact between the two groups was made and maintained. A vital meeting for all concerned and for the world at large, was that which was

Addendum

held in London on 28th of November 1660. The group had met at Gresham College to hear a lecture by Sir Christopher Wren, and following their usual custom, adjourned after the lecture for discussion and conversation. During the conversation the suggestion was made and approved that the group should be more formalized into a Society for the study of the new philosophy, with regular membership based on qualifications. The New Society was actually formed and instituted, and minutes of proceedings taken with a view to publication. Soon after the inception the king indicated his willingness to be enrolled as a member, and after his acceptance into the Society, he offered it a charter of incorporation, which was granted in 1662, thus constituting it the Royal Society. The membership was limited to a small number, fifty-five in the earliest years, which included among the foundation members many eminent names ... at least two* of which who were closely linked with important Friends.

During the first forty years of the Royal Society, Friends by their exclusion from the Universities and professions, had little chance of attaining the qualifications for a fellowship, but nonetheless some Friends had close associations with Fellows. In the eighteenth century Friends through their outstanding work in medicine, natural history, and instrument making soon secured election as Fellows and occupy an honourable position in its role. Many Friends who were not Fellows were intimate with leading scientists and this awareness of the progress of scientific thought and the frequent discussions of developing theories must have had weight in the formation of the intellectual background of the Society.

---

*Sir John Finch and Anthony Lowther. See *Friends Hist. Journal* VII, 1910, p. 30. Sir John Finch was the brother of Lady Anne Conway, the Quakeress. Anthony Lowther was the brother-in-law of William Penn.

Addendum

From *Friends for 300 Years* by Howard Brinton. Harper Brothers, New York, 1952

The search for an understanding of the creation and insight into its beauty, sincerity and genuineness led many Quakers to scientific pursuits, particularly, botany and ornithology. Some became professional scientists. Science seemed closer to reality than did art. With the exclusion of many forms of amusement, it also afforded delight. Superfluities in education were eliminated as completely as other nonessentials. Jonathan Dymond, the Quaker moralist, writes in 1825: "Science is preferable to literature, the knowledge of things to the knowledge of words."

Verbalism and formalism were opposed in education as they were in religion. Knowledge of God's creation was thought to bring man nearer to the divine than a knowledge of man's works. To quote Dymond again:

> *It is of less consequence to a man to know what Horace wrote or to be able to criticize the Greek anthology than to know by what laws the Deity regulates the operations of nature and to know by what means those operations are made subservient to the purposes of life.*

As a consequence of this scientific interest, which was a direct result of the effort to come closer to sincerity and reality, the list of Quaker scientists is a long one. A. Ruth Fry observes in *Quaker Ways* that between 1851 and 1900 in England a Quaker "had forty-six times more chance of election as a Fellow of the Royal Society than his fellow countryman."

Addendum

From *Science and the Unseen World* by Sir Arthur Stanley Eddington. George, Allen & Unwin: London, 1929

The spirit of Seeking in science and religion

In its early days our Society [the Religious Society of Friends] owed much to a people who called themselves Seekers; they joined us in great numbers and were prominent in the spread of Quakerism. It is a name that must appeal strongly to the scientific temperament. The name has died out, but I think the spirit of seeking is still the prevailing one in our faith, which for that reason is not embodied in any creed or formula. It is perhaps difficult sufficiently to emphasise Seeking without disparaging its correlative Finding. But I must risk this, for Finding has a clamorous voice that proclaims it own importance; it is definite, assured, something we can take hold of – that is what we all want, or think we want. Yet how transitory it proves. The finding of one generation will not serve for the next. It tarnishes rapidly except it be preserved with an ever-renewed spirit of seeking. It is the same too in science. How easy in a popular lecture to tell of the findings, the new discoveries which will be amended, contradicted, superseded in the next fifty years! How difficult to convey the scientific spirit of seeking which fulfills itself in this tortuous course of progress toward truth! You will understand the true spirit of neither science nor religion unless seeking is placed in the forefront.

Religious creeds are a great obstacle to any full sympathy between the outlook of the scientist and the outlook which religion is so often supposed to require. I recognise that the

Addendum

practice of a religious community cannot be regulated solely in the interests of its scientifically-minded members and therefore I would not go so far as to urge that no kind of defense of creeds is possible. But I think it may be said that Quakerism in dispensing with creeds holds out a hand to the scientists. The scientific objection is not merely to particular creeds which assert in outworn phraseology beliefs which are either no longer held or no longer convey inspiration to life. The spirit of seeking which animates us refuses to regard any kind of creed as its goal. It would be a shock to come across a university where it was the practice of the students to recite adherence to Newton's laws of motion, to Maxwell's equations and to the electro-magnetic theory of light. We should not deplore it the less if our own pet theory happened to be included, or if the list were brought up to date every few years. We should say that the students cannot possibly realise the intention of scientific training if they are taught to look on these results as things to be recited and subscribed to. Science may fall short of its ideal, and although the peril scarcely takes extreme form, it is not always easy, particularly in popular science, to maintain our stand against creed and dogma. I would not be sorry to borrow for scientific pronouncements the passage prefixed to the Advices of the Society of Friends in 1656 and repeated in the current General Advices:

> *These things we do not lay upon you as a rule or form to walk by; but that all with measure of the light, which is pure and holy, may be guided; and so in the light walking and abiding, these things may be fulfilled in the Spirit, not in the letter; for the letter killeth, but the Spirit giveth life.*

Rejection of creed is not inconsistent with being possessed by a living belief. We have no creed in science, but we are not lukewarm in our beliefs. The belief is not that all the knowledge of the universe that we hold so enthusiastically will survive in the letter; but a sureness that we are on the road. If our so-called facts are changing shadows, they are shadows cast by the light of constant truth. So too in religion we are repelled by that confident theological doctrine which has settled for all generations just how the spiritual world is worked; but we need not turn aside from the measure of light that comes into our experience showing us a Way through the unseen world.

Religion for the conscientious seeker is not all a matter of doubt and self-questionings. There is kind of sureness which is very different from cocksureness.

References

Frontispiece Quotations

1. Berry, Wendell, 1995. *Another Turn of the Crank.* Washington: Counterpoint.

2. Hopkins, Gerard Manley, 1954. *Poems and Prose.* London: Penguin Books.

3. Schwenk, Theodor, 1976. *Sensitive Chaos: The Creation of Flowing Forms in Water & Air.* New York: Schocken Books.

4. Coffin, William Sloan, 2004. *Credo.* Louisville KY: Westminster John Knox Press.

Introduction: From Theology to Continuing Revelation

1. http://en.wikipedia.org/wiki/Mathematics

2. http://www.lyricstime.com/raffi-the-cat-came-back-lyrics.html

References

3. http://www.brainyquote.com/citation/quotes/quotes/p/ paultillic154766.html#aRz8eOCx8Xi4TDJj.99

4. http://www.goodreads.com/quotes/8703-he-drew-a-circle-that-shut-me-out--heretic

5. Fox, George, 1998. *The Journal*. London: Penguin Books.

6. Boulding, Kenneth, 1964. *The Evolutionary Potential of Quakerism*. Wallingford PA: Pendle Hill

From Theism to a Kind of Pantheism

1. Haught, John, 1995. *Science and Religion: From Conflict to Conversation*. New York: Paulist Press.
See also: Popper, Karl, 2004. *Conjectures and Refutations: The Growth of Scientific Knowledge*. London: Routledge.

2. Harpur, Tom, 2004. *The Pagan Christ*. Toronto: Thomas Allen.

3. Dawkins, Richard, 2008. *The God Delusion*. New York: Houghton Mifflin.

4. Swimme, Brian and Thomas Berry, 1992. *The Universe Story: From the Primordial Flaring Forth to the Ecozoic Era; A Celebration of the Unfolding of the Cosmos*. New York: HarperCollins.

5. Churchill, Caryl, 2003. *A Number*. New York: Theatre Communications Group.

6. http://en.wikipedia.org/wiki/Pantheism

7. Isaacson, Walter, 2008. *Einstein: His Life and Universe*. New York: Simon and Schuster.

8. Isaacson, ibid.

9. Dawkins, op cit.

10. http://www.pantheism.net

11. Dawkins, op cit.

12. Weinberg, Stephen, 1993. *Dreams of a Final Theory*. London: Vintage.

13. Dennett, Daniel C., 1991 *Consciousness Explained*. Boston: Little, Brown

14. Dawkins, op cit.

15. Russell, Sharman Apt, 2008. *Standing in the Light: My Life as a Pantheist*. New York: Basic Books.

16. Bouton, David, Editor, 2006. *Godless for God's Sake: Nontheism in Contemporary Quakerism*. Dent, Cumbria, UK: Dales Historical Monographs.

Additional Recommended Reading

Runes, Dagobert D., 1995. *The Ethics of Spinoza*. New York: Carol Publishing Group.

References

Imagination and Belief

1. Greene, Graham, 1940. *The Power and the Glory*. London: Penguin Books.

2. Faber, Frederic W., 1814-1863. "Faith of Our Fathers." www.hymnsite.com

3. Raymo, Chet, 1998. *Skeptics and Believers: The Exhilarating Connection Between Science and Religion*. New York: Walker and Company.

4. *American Heritage Dictionary of the English Language*, 2004. Boston: Houghton Mifflin Harcourt.

5. Ancelet-Hustache, Jeanne, 1957. *Master Eckhart and the Rhineland Mystics*. London and New York: Longmans, Green and Company, Harper and Brothers.

6. Boulding, Kenneth E., 1956. *The Image: Knowledge in Life and Society*. Ann Arbor MI: University of Michigan Press.

7. *American Heritage Dictionary*, op. sit.

8. Lonergan, Anne and Caroline Richards, editors, 1988. *Thomas Berry and the New Cosmology*. Mystic CN: Twenty-Third Publications.

9. *The Bible: New Testament*, King James Version. "Book of Matthew," chapter 23.

10. Ibid. "Book of John," chapter 18:36.

11. Voltaire, 2003 (1759). *Candide*. New York: Barnes and Noble.

12. The Bible, op. sit. "Book of John," chapter 4:20.

13. Wright, Robert, 2009. *The Evolution of God: The Origins of Our Beliefs*. New York: Little, Brown and Company.

14. Newberg, Andrew, 2006. *Why We Believe What We Believe: Uncovering Our Biological Need for Meaning, Spirituality, and Truth*. New York: The Free Press.

15. Ridley, Matt, 1998. *The Origins of Virtue: Human Instincts and the Evolution of Cooperation*. New York: Penguin Books.

16. Wright, Robert, 2000. *Nonzero: The Logic of Human Destiny*. New York: Pantheon Books.

17. Harris, Sam, 2004. *The End of Faith: Religion, Terror, and the Future of Reason*. New York: W.W Norton.

18. Dawkins, Richard, 2006. *The God Delusion*. New York: Bantam Books.

19. Hitchens, Christopher, 2007. *God Is Not Great: How Religion Poisons Everything*. New York: Twelve Books.

References

Faith Behind Faith

1. Wissler, Clark, 1926. *The Relation of Nature to Man in Aboriginal America*. New York: Oxford University Press.

2. Sauer, Carl Ortwin, 1967. *Land and Life*. Berkeley CA: University of California Press.

3. Teilhard de Chardin, Pierre, 1970. *Activation Energy*. London: Collins.

4. Blakney, Raymond, trans., 1957. *Meister Eckhart: The Essential Writings*. New York: Harper

5. Lovelock, James E., 1979. *Gaia: A New Look at Life on Earth*. New York: Oxford University Press.

6. Berry, Thomas, 1988. *The Dream of Earth*. San Francisco: Sierra Club Books.

7. Berry, Thomas and Brian Swimme, 1992. *The Universe Story: From the Primordial Flaring Forth to the Ecozoic Era; A Celebration of the Unfolding of the Cosmos*. San Francisco: HarperCollins.

8. Berry, Thomas, 1999. *The Great Work: Our Way Into the Future*. New York: Bell Tower Books.

9. Hocking, Brian, 1965. *Biology or Oblivion: Lessons from the Ultimate Science*. Cambridge MA: Scheckman.

10. Commoner, Barry, 1971. *The Closing Circle: Man, Nature, and Technology*. New York: Alfred Knopf.

11. Commoner, Barry, 1976. *The Poverty of Power: Energy and the Economic Crisis*. New York: Alfred Knopf.

12. Leopold, Aldo, 1966. *A Sand County Almanac*. New York: Oxford University Press.

13. Fox, George, 1924. *Journal of George Fox*. London: J. M. Dent.

14. Cobb, Edith, 1977. *The Ecology of Imagination in Childhood*. New York: Columbia University Press.

15. Brown, Peter G., Geoffrey Garver, Keith Helmuth, Robert, Howell, and Steve Szeghi, 2009. *Right Relationship: Building a Whole Earth Economy*. San Francisco: Berrett-Koehler.

16. Boulding, Kenneth, 1966. "The Economics of the Coming Spaceship Earth." http://www.eoearth.org/view/article/156525/ Kenneth Boulding is also famous for saying, "Anyone who believes that exponential growth can go on forever in a finite world is either a madman or an economist." http://www.goodreads.com/quotes/627148-anyone-who-believes-that-exponential-growth-can-go-on-forever

Acknowledgements

Formative influences at a young age seem to gain resonance in the latter part of life. As Viktor Frankl says, "Memories are real possessions." (*Man's Search for Meaning*). Accordingly, our acknowledgements honour significant relationships from our early years.

George Strunz wishes to record his appreciation to the Friends of Dublin Monthly Meeting and Churchtown Meeting in Ireland as well as Newtown School, Waterford, Ireland, who in the 1940s and 50s provided him with a nurturing and supportive Quaker environment in which to grow during his early years.

Michael Miller wishes to acknowledge his parents, Stephen and Lena, and note with gratitude their open family discussions around the dinner table on religion and science. He thanks his twin sister, Doreen, for putting up with his antics in their early years, and his brother Tony for frank exchanges on God and human nature in more recent times. He also wishes to acknowledge Brooks Ellis, a geology professor at New York University for enlarging his worldview to include the last billion years of Earth history.

Acknowledgements

Keith Helmuth wishes to record his appreciation for his parents, Atlee and Naomi, who had a spirit of kindliness and an attitude of service that made religious and daily life one reality. He is also grateful to Dr. Y.P. Mei, a gracious Confucian scholar at the State University of Iowa, for his mentoring in Asian studies, and to Murray Bookchin for guidance in opening the field of social ecology. Special acknowledgement also goes to Morris Mitchell, founding president of Friends World College, and Arthur Meyer, Director of the North American Center, for the opportunity to help develop the College's experiential world education program.

We also wish to express special appreciation to Edith Miller and Ellen Helmuth for their editing and proofreading work, and to Brendan Helmuth for his design and layout of the book.

About the Authors

George M. Strunz

George Strunz was born in Vienna, Austria, but moved as an infant with his family to Ireland on the eve of World War Two. The family became members of the Religious Society of Friends (Quakers) soon after their arrival in Ireland. He received his high school education at Newtown, a Quaker boarding school in Waterford. Following his undergraduate degree in Natural Sciences from Trinity College, Dublin, he moved to Canada for graduate studies in organic chemistry at the University of New Brunswick. Subsequently, he held postdoctoral fellowships at the University of Michigan and Harvard University before returning to Fredericton, New Brunswick where he has since resided.

For most of his professional career he worked as a Research Scientist with the Canadian Forest Service and as Adjunct Professor of Chemistry at the University of New Brunswick. His scientific research focused on the chemistry of natural products, primarily in the context of their potential role in maintaining the health of the forest. He has published extensively on this and related topics.

About the Authors

He retired in 1999, which has given him the opportunity to devote time to other interests, especially the enjoyment of five grandchildren. George and his wife, Annette, have two married daughters. He is a veteran member of the Fredericton Choral Society. Painting, a long time hobby, has become an important part of his life and he has held several solo art exhibitions. He enjoys outdoor recreational activities, including skiing and snowshoeing in winter and kayaking in summer.

He has been a Board member of professional and volunteer community organizations, including the New Brunswick Choral Federation, Fredericton Arts Alliance, and the Fredericton Botanic Garden. He is a member of the New Brunswick Monthly Meeting of the Religious Society of Friends (Quakers). George and Annette spend summers at their home in Queenstown on the Lower Saint John River, and winters in Fredericton.

Michael R. Miller

Michael R. Miller was born in 1932 in Portugal where he lived with his family until 1941. He then lived in and around New York City, except for nine months when his father's employment took the family to Lagos, Nigeria. He holds a B.A. from New York University and an M.A. and Ph.D. from the Eastman School of Music in Rochester, NY. He married Edith Hoisington in 1962. With their three sons they moved to Sackville, NB in 1967 where he became Professor of Music at Mount Allison University.

Michael is a composer and pianist. He taught at Mount Allison University until retirement in 1999, at which time he and Edith moved to Fredericton, NB. His many compositions include works for voice, choir, orchestra, and various chamber groups.

About the Authors

His most recent work, Marsh-Boy Music, is based on a selection of Douglas Lochhead's poems from the book *High Marsh Road*. A Creation Grant from ArtsNB supported this composition. A new chamber group, Atlantica, premiered Marsh-Boy Music in September 2013

Michael and Edith became members of the Religious Society of Friends (Quakers) in 1975 while on sabbatical leave in Montreal. He served as the first Clerk of New Brunswick Friends Meeting when it was established in 1980. He is on the Boards of Symphony New Brunswick and the John Howard Society, and is a member of the Spiritual and Religious Care Committee of the Doctor Everett Chalmers Hospital in Fredericton. He represented Quakers at the "Celebration of Faith in Diversity" at St. Thomas University in 2010. He has published some of his writings on faith and religion in Canadian Friend.

Michael and Edith enjoy frequent visits with their grandchildren. They enjoy birding and summer retreats on Edith's family island off the coast of Maine. Their three sons are all professional musicians. Michael and Edith are members of the New Brunswick Monthly Meeting of the Religious Society of Friends (Quakers).

Keith Helmuth

Keith Helmuth was born in Ohio in 1937 where he grew up working with his father in the family roofing business and with his mother in the family garden. After graduating from the State University of Iowa, he managed academic bookstores in Iowa City, Iowa, Syracuse, New York, and New York City. In 1967 he joined the faculty of Friends World College at the

About the Authors

North American campus on Long Island, New York. He helped establish the College's Independent Studies Program and served as its first Coordinator. He and his wife, Ellen, also worked with the College's programme in East Africa.

From 1972 through the late 1990s, Keith and Ellen, along with their two sons, operated a farm and market garden business in the St. John River Valley of New Brunswick. During this time he also worked with community economic development projects, including farm markets, credit unions, a gristmill cooperative, and an employment training agency. In 1991, Keith was the Canadian Quaker delegate to the World Council of Churches' Convocation on Justice, Peace and the Integrity of Creation in Seoul, Korea. He is a co-author of two books: *Right Relationship: Building a Whole Earth Economy*, and *How on Earth Do We Live Now? Natural Capital, Deep Ecology, and the Commons*. He has been publishing articles and making presentations on the intersection of ecology, economics, ethics, and religion for over four decades.

After retiring from farming in 1998, Keith returned to the bookstore business as manager of Penn Book Center on the campus of the University of Pennsylvania in Philadelphia. He was instrumental in founding Quaker Institute for the Future (2003) and served as its first Board Secretary and Coordinator of Publications. He served in an advisory capacity with Quaker International Affairs Programme (Ottawa) from 2005 to 2010. In 2008, he and his wife Ellen returned to New Brunswick, where they are now on the Board of the Woodstock Farm Market Co-operative and work with the Transition Town Woodstock initiative. They are active in community gardening and are members of New Brunswick Monthly Meeting of the Religious Society of Friends (Quakers).

About the Publisher

Chapel Street Editions publishes fine books on the natural history, human history, and cultural life of the Saint John River Region.

We are dedicated to publishing the work of writers and artists of our region, and to publishing books that advance an understanding of the relationship between the natural world, culture, and human adaptation to place.

We believe a vibrant cultural life rests on a strong attachment to place: This means a strong attachment to dwelling with the land, the built environment, and the communities where we live.

For additional information visit
www.chapelstreeteditions.com

www.ingramcontent.com/pod-product-compliance
Lightning Source LLC
Chambersburg PA
CBHW032043290426
44110CB00012B/933